Geographic Information Systems: Design, Technology and Applications

Geographic Information Systems: Design, Technology and Applications

Edited by **Marina De Lima**

LANRYE
INTERNATIONAL

New Jersey

Published by Clanrye International,
55 Van Reypen Street,
Jersey City, NJ 07306, USA
www.clanryeinternational.com

Geographic Information Systems: Design, Technology and Applications
Edited by Marina De Lima

International Standard Book Number: 978-1-63240-558-6 (Hardback)

Printed in the United States of America.

Contents

Preface

Every book is a source of knowledge and this one is no exception. The idea that led to the conceptualization of this book was the fact that the world is advancing rapidly; which makes it crucial to document the progress in every field. I am aware that a lot of data is already available, yet, there is a lot more to learn. Hence, I accepted the responsibility of editing this book and contributing my knowledge to the community.

Geographic information systems are designed for the purpose of mapping any physical location on earth. They are designed to store, capture, analyze and present different forms of spatial data. It is applicable to various areas such as climatology, landscape architecture, natural resources, crime mapping, real estate, archaeology, public health and many others. The ever growing need for advanced technology is the reason that has fueled the research in this field in recent times. The objective of this book is to give a general view of the different areas of geographic information system and its applications. For all readers who are interested in this area, the case studies included in this book will serve as an excellent guide to develop a comprehensive understanding.

While editing this book, I had multiple visions for it. Then I finally narrowed down to make every chapter a sole standing text explaining a particular topic, so that they can be used independently. However, the umbrella subject sinews them into a common theme. This makes the book a unique platform of knowledge.

I would like to give the major credit of this book to the experts from every corner of the world, who took the time to share their expertise with us. Also, I owe the completion of this book to the never-ending support of my family, who supported me throughout the project.

Editor

GIS and Networks: Business Anomalies and Topological Errors, Linear Elements Case

Omar Bachir Alami[1], Hatim Lechgar[2], Mohamed El Imame Malaainine[2], Fatima Bardellile[1]

[1]Ecole Hassania des Travaux Publics, Casablanca, Maroc
[2]Faculté des Sciences Ain Chock, Université Hassan II, Casablanca, Maroc
Email: alami.ehtp@gmail.com, h.lechgar@gmail.com, mohamadlimam@gmail.com, fatima.bardellile@gmail.com

Abstract

The quality control of geographic data, especially from a topological and semantic perspective, is a must for its good management and use. However, while updating spatial data, some sorts of anomalies are affecting it, due to negligence or non-respect of business and topological rules. Hence the necessity of a solution that enables detecting theses anomalies. Nowadays, Geographic Information Systems (GIS) have become essential for decision-making in any project that manages spatial data. GIS functionalities and tools give the possibility of defining the topology of vector data. Nevertheless, the topology alone does not respond to the needs in matter of defining specific rules for every facility network. This means, we could find topological errors in the spatial database, but taking into account business rules, they are correct and *vice versa*. The main objective of this article is firstly to define business rules for the linear elements of a network. Secondly to premeditate the algorithms that detect the violation of the defined rules in order to have a good quality control of geographic data.

Keywords

GIS; Networks, Geodatabase, Topology, Business Rule

1. Introduction

GIS are ever evolving and offer more sophisticated analysis functionalities. To keep up with this evolution, geographic data ought to be exploitable and of good quality. The term quality is defined, according to ISO 8402 [1] norm as "the totality of features and characteristics of a product or service that bears its ability to satisfy stated or implied needs". For a geographic database it is rather delicate to separate the product (data base) from services (updating, extraction of certain themes, and conversion to certain data format) [2].

Thus, this definition of quality has been adapted to the specific domain of geographic information, while splitting the quality in two parts:

- The intern quality which consists of measuring the adequacy of database to its specifications.
- The extern quality which consists of measuring the adequacy to the users' needs.

The creation or update of Geodatabase (GDB) might be source of several anomalies which are due to human omission or to non respect of topological or business rules, hence the interest of having a tool that would control the quality of GDB. In fact, every advanced GIS functionality requires a high standard GDB. For example, to determine the impact zones during electricity power failure, and intervene as quickly as possible, the electric network has to be completely connected and respects the topological and business rules. Nevertheless, we might have data with topological errors in GDB but taking into account business criteria those data are correct, or *vice versa*.

The verification of geographic data following the standard procedures offered by GIS software, which by the way does not take into consideration business rules, turns out to be limited, repetitive and tedious. Whence the need of proposing algorithms is to detect errors and anomalies, in particular topological and business ones which are the most recurrent in GDBs for linear elements in a utility network.

2. Methodology

The present work is based on a dataset of elements from utilities network, namely: canalizations of potable water and waste water networks, high voltage (HV) and low voltage (LV) networks sections.

Our methodological approach consisted firstly of the study of the relationships between linear GDB entities to determine the most common business rules in utilities network, and secondly of the development of algorithms to detect anomalies which violate these rules.

In order to detect the violation of a predefined rule, all concerned entities have to be scanned. Nevertheless, we are dealing with huge quantities of geographic data and the spatial queries tends to be rather complex. So, a linear scan of the GDB records to spot nonconformities to a rule takes a lot of time. That is why we adopted the following general method in order to accelerate anomalies detection, and thus optimize search time.

The first step, which is filtering, aims to cast aside all entities whose attributes does not fit the controlled business rule criteria, then throw out all remaining entities that are not close to the controlled entity. The next step is refinement, in which the exact search for violations takes place so that entities subjects of anomalies are detected (**Figure 1**).

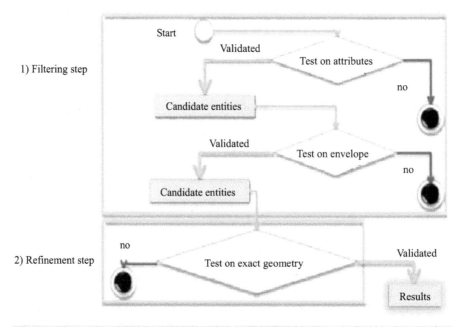

Figure 1. Processes of filtering and refinement of spatial queries [3].

3. Results

3.1. Rules for Linear Elements

A detailed study of the relationships between these entities in the GDB has allowed us to define these generalizing rules:

3.1.1. Rule 1: Must Be Connected

This rule ensures that two polylines whose ends are close by a distance "d" must be connected. "d" is a margin specified by the quality controller while defining this rule (**Figure 2**).

3.1.2. Rule 2: Must Be Connected to

In this rule, two polylines of two different feature classes whose ends are close by a distance "d" must be connected. "d" is a margin specified by the quality controller while defining this rule. For instance; a low voltage power cable must be connected to a cable of the same type or of medium voltage power and a combined sewer pipe must be connected to a storm water sewerage pipe.

3.1.3. Rule 3: Must Have the two Extremities on

In this rule, the two ends of a polyline must be on other polylines. For example; two ends of an electric cable must be above a connecting node and both ends of potable water canalization, which must be above a node.

3.1.4. Rule 4: Should Not Have an End near the Inside of

This rule requires that the end of a polyline must not be close by a distance "d" to the inside of another polyline as with the case of a drinking water or wastewater pipe which should never touch the inside of another pipe (**Figure 3**).

3.1.5. Rule 5: Should Not Be Connected to

This rule ensures that the end of a polyline must not coincide with the end of another polyline. This is the case of a rainwater pipe which should not be connected to waste water pipe.

3.1.6. Rule 6: Must Not Overlap

In this rule, the segments of a line should not occupy the same location with different line segments. This is the case of duplicate drinking water pipes, sewerage or electrical cables. One element must be deleted. The following

Figure 2. Anomaly example: one end of a low voltage electrical line is 9 cm close to the ends of other power lines but are not connected.

Figure 4 shows the overlapping anomaly of two pipes of potable water (successively selected in cyan blue): the two circles locate the overlapping.

3.1.7. Rule 7: Must Not Self Overlap

In this rule, the segments of a polyline must not occupy the same location with segments of the same polyline. A drinking water pipe should never self overlap.

3.1.8. Rule 8: Should Not Cross

This rule requires that a polyline should not cross another polyline. This is to check for the three businesses (water, sanitation and electricity), save the case when it is just a warning and not a systematic error, since some pipes with different elevations may appear intersecting. Z is to take into account in the detection algorithm of this rule.

3.1.9. Rule 9: Should Not Self Cross

In this rule, a segment of a polyline must not cross another segment of the same polyline. This is applicable on a drinking water and sanitation pipes (**Figure 5**).

The following **Table 1** summarizes all the business rules studied.

3.2. Algorithms for Detecting Anomalies on Linear Elements

In this section we detail the algorithms developed to find violations of defined topological constraints. We opted to use the intersection of geographical entities envelopes instead of their exact geometry because it is a function managed intrinsically and effectively by GDBs.

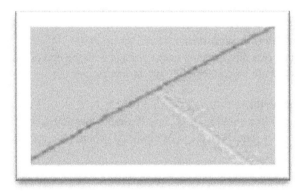

Figure 3. Anomaly example: the end of an electric line is close to the inside of another power line.

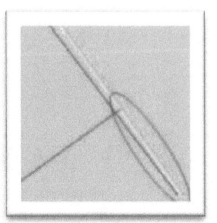

Figure 4. Anomaly example: overlapping of two polylines in cyan.

Figure 5. Anomaly example: a drinking water pipe that self crosses.

Table 1. List of rules between the linear elements.

Rule number	Rules between linear elements
1	Must be connected
2	Must be connected to
3	Must have the two extremities on
4	Should not have an end near the inside of
5	Should not be connected to
6	Must not overlap
7	Must not self overlap
8	Should not cross
9	Should not self cross

3.2.1. Detection Algorithm of Rules 1 and 2

For the rule "Must be connected":

For every polyline whose entity class is defined in the rule, we create a buffer disk of radius "d" around its first end. We restrict the search to polylines whose envelope intersects the envelope of buffer as illustrated in the following **Figure 6**.

Three cases are then distinguished:

1) The envelope of a polyline intersects the buffer's envelope without having any of its two ends belonging to the buffer (**Figure 7**).

→ Rejected case.

2) The polyline envelope intersects the buffer's envelope, and one of its ends is contained in the buffer belongs while coinciding with the end of the controlled polyline (**Figure 8**).

→ Rejected case (no anomaly)

3) The polyline envelope intersects the buffer's envelope, and one of its ends is within the buffer without coinciding with the end of the first controlled polyline (**Figure 9**).

→ Case detected as anomaly.

The same procedure is followed for the second end of the polyline (endpoint). Identifying the types of ends (end or start) involved in the detected anomaly is essential for the visualization and correction of the anomaly.

For rule 2, the detection algorithm is the same as that of Rule 1, except that in this case polylines do not belong to the same class of entities.

3.2.2. Detection Algorithm of Rule 3

A polyline is considered a problem if one of its two ends (startpoint or endpoint) is not above another polyline. To detect these anomalies, we followed an indirect method by omitting from the group of all polylines, those who have at least one end over another polyline (which shows to be time effective compared to the direct approach,

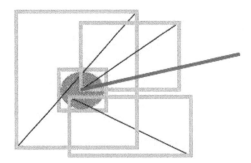

Figure 6. The controlled polyline in pink and its created buffer in red.

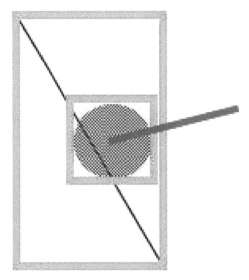

Figure 7. Illustration of the first case.

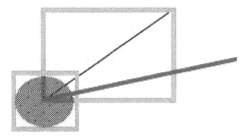

Figure 8. Illustration of the second case.

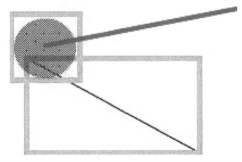

Figure 9. Illustration of the third case.

knowingly: searching for polylines whose start and end points intersect no polyline).

For this we look at first for the group of all polylines whose start point is at the top of another polyline then we search the group of all polylines with the End point above another polyline. So the Polylines source of anomaly are those who belong neither to the first group nor to the second group.

- Method to find the first group

To find the first group we seek polylines whose end startpoint is above another polyline. To do this, we calculate for each polyline the distance "d" between the startpoint and other polylines, from the second class of entities, whose envelope intersects the envelope of the polyline in question (**Figure 10**).

So the first group is the set of polylines with distance d = 0.

- Method to find the second group

Then we look for polylines whose endpoint is above another polyline in the same manner as for the startpoint. Therefor group 2 is the set of polylines with a distance "d" of zero. Polylines considered anomalies are then those not belonging to either group1 or to group 2.

3.2.3. Detection Algorithm of Rule 4

We create a buffer disk of radius "d" around an end of the polyline subject to rule control. Thereafter, the search will be restricted to polylines intersecting with the buffer (**Figure 11**).

The following cases are to get rid of:

- The end of the polyline in question touches another polyline result of the first filter, at its inside or at its end.
- The extremity of the other polyline is in the buffer, since this case is treated in interconnection anomaly detecting.

3.2.4. Detection Algorithm of Rule 5

We create a buffer disk of radius "d" around the end (start point) of the polyline to be examined. Next, we check whether the start point coincides with an end of another polyline (We are interested in polylines intersecting with the already established buffer). Polylines considered as anomaly are those with the "start point" or "end point" which coincides with the end of another polyline.

3.2.5. Detection Algorithm Rules 6 and 7

For the rule "should not overlap", we seek for each polyline the polylines overlapping it among those whose envelope is interesting its envelope. For the rule "should not self overlap," the same algorithm as the rule "should not overlap" is applied, except that in this case we are interested in the relationship of the same polyline.

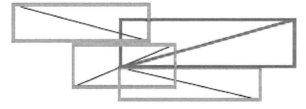

Figure 10. The polyline in question and its envelope are in red.

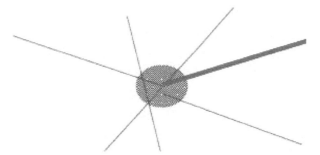

Figure 11. The controlled polyline in pink and its created buffer in red.

3.2.6. Detection Algorithm Rules 8 and 9

For the rule "should not cross" for each polyline we look for crossing polylines from those having the envelope intersecting with its envelope. For the rule "should not self cross" we apply the same algorithm of the previous rule except that this is about the same polyline.

4. Conclusions

In conclusion, we have proposed the most widely used business rules in the field of network utilities and algorithms for detecting violations of these rules on linear elements of these networks (drinking water, sanitation and electricity).

This work is intended for the use of quality controllers and business agents in order to define business rules on their networks. It is also intended to detect report and correct anomalies in order to have more clean and reliable geographic data which is an essential precondition to any spatial analysis or GIS application.

It should be noted that available GIS solutions do not allow meeting all and each business's set of rules, which is why it is necessary for companies that manage network utilities to develop according their specific needs, solutions for the detection and automatic or semi-automatic correction of their business anomalies and that, based on our algorithms and others.

References

[1]   ISO (1994) International Organization for Standardization, Norme Internationale—Management de la qualité et assurance qualité—Vocabulaire. ISO 8402:1994 (E/F/R), 2ème édition.

[2]   Bonin, O. (2002) Modèle d'erreurs dans une base de données géographiques et grandes déviations pour des sommes pondérées; application à l'estimation d'erreurs sur un temps de parcours. Thèse de doctorat, Université de Paris VI, 147.

[3]   Papadopoulos, A.N. and Manolopoulos, Y. (2005) Nearest Neighbor Search: A Database Perspective. Series in Computer Science. Springer, 170.

2

Spatial Modeling of Optimum Zones for Wind Farms Using Remote Sensing and Geographic Information System, Application in the Red Sea, Egypt

Hala A. Effat

Department of Environmental Studies and Land Use, National Authority for Remote Sensing and Space Sciences (NARSS), Cairo, Egypt
Email: heffat@narss.sci.eg, haeffat@yahoo.com

Abstract

Wind power is a safe form of renewable energy and is one of the most promising alternative energy sources. Worldwide, the wind power industry has been rapidly growing recently. It is crucial that the locating of new projects must address both environmental and social concerns. The Red Sea shoreline in Egypt provides excellent wind power potential sites for the Red Sea Governorate. In this study, appropriate zones for wind power farms were mapped using remotely sensed data and a GIS-based model namely Spatial Multi-Criteria Evaluation (SMCE). This model incorporated several criteria, two sets of factors and a set of constraints. First, resource factors included wind speed, elevation zones used to derive the wind power density. Second, economic factors included distances from urban areas, roads and power-lines. Third, land constraints were excluded from the evaluation. The land constraints set included land slope angles, shoreline, urban areas, protectorates airports and ecologically sensitive and historical areas. The Analytical Hierarchy Process was used to assign the criteria relative weights. The weighted criteria and constraints maps were combined in the MCE model. The model identified the zones with potential wind power energy. Such zones were found to exist along the northern parts of the Red Sea shoreline. Some of which are unsuitable due to their location within a sensitive eco-system, high slopes and/or a nearby airport. By excluding such land constrains, the model identified the most appropriate zones satisfying all assigned suitability conditions for wind farms. Ideal zones amount to 706 sq. km with suitability values ranging from 83% to 100% and highly suitable zones amount to 3781 sq. km having suitability values ranging from 66% to 83%.

Keywords

Wind Power, Remote sensing, GIS, Multicriteria, Red Sea, Egypt

1. Introduction

1.1. General

An increase in public awareness regarding the negative impact of traditional power-generating methods, especially coal and oil-fired power stations, has created a demand for using environmentally friendly renewable energy sources. Developing electrical energy from renewable sources is becoming a necessity because it does not release harmful emissions into the environment (Fernandez *et al.*, 2006) [1].

Historically, wind power has been used as a source of power for ships and wind mills. Today, huge wind turbines are used as a renewable energy resource for generating electricity. They are erected in particularly windy places, because wind farms can operate only where steady winds prevail year round.

Various environmental impacts of wind energy are commonly known by scientists. These impacts may be listed as effects on animal habitats (particularly bird collisions), noise generation, visual impact, and electromagnetic interference. A further impact of wind energy on habitat is noise. Example of regulations by authorities as given by Tester *at al.* (2005) [2]; Ramiraz-Rosado *et al.* (2008) [3] stated that wind turbines should be located at least 500 m away from nearest habitat and they studied the visual impact of the wind turbines. Baban and Parry (2001) [4], Nguyen (2007) [5] stated that wind turbines should be located 2000 m away from large settlements because of aesthetic concerns and safety. According to Yue and Wang (2006) [6] wind turbines must be located at least 500 m away from wildlife conservation areas and ecological sites such as bird migration sites.

Site selection for large wind turbine requires consideration of a comprehensive set of factors and balancing of multiple objectives in determining the suitability of a particular area for a defined land use (Bennui *et al.*, 2007) [7]. Literature on the siting of wind power facilities generally incorporates the integration of many factors in order to determine a suitable location. Wind resources are the most important criteria. Such resource is classified based on the annual mean wind speed (Patel and Rosier, 2013) [8]. For annual average wind speed between 6 and 10 m/s its production varies between 2.4 and 6.5 GWh. Wind farms require a lot of space. Most wind farms fall into a range of 0.1 - 1 km^2 per installed MW (Elliason, 1998; Walker and Jankens, 1997; Afgan and Carvalho, 2002) [9]-[11].

GIS can have significant contribution as a decision support tool in identifying environmentally feasible locations for wind turbines which require management and analysis of wide range of spatial data types. GIS analysis might aid to determine appropriate zones according to specific criteria for future development. A MCE is the evaluation of a set of alternatives, based on multiple factors and constraints, where the factors are quantifiable indicators of the extent to which decision objectives are realized (Malczewski, 1999) [12]. The most important factor in MCE is how to establish "weights" for a set of criteria according to importance. Location decisions such as the ranking of alternative communities are representative multi-criteria decisions that require prioritizing multiple criteria. A MCE enables the outcomes to be visualized in maps (Wood and Dragicevic, 2007) [13]. One of the most useful GIS decision making tools is the Analytical Hierarchy Process (AHP). This process is a comprehensive, logical and structural framework, which allows analyzer to improve the understanding of complex decisions by decomposing the problem in a hierarchical structure. Such structure shows the relationship between the goal, objectives, criteria, and alternatives (Bennui *et al.*, 2007) [7]. This will also allow for unique ecological, social, economic and/or combined scenarios to be modeled (Mann *et al.*, 2012; Bartnicki and Williamson, 2012) [14] [15]. This process enables the analyst to perform multiple scenarios by altering factor weights depending on the user's preference.

Different modeling approaches have been conducted to arrive to the most appropriate sites for wind farms. Economic factors for the siting of wind power facilities include wind speed and/or wind power density and distance to hydro lines, roads and slopes. Socially influenced factors included distance from urban areas and historic sites. Environmental variables generally included distance from wetlands, forests and water bodies (Baban and Perry, 2001; Mann *et al.*, 2012; Rodman and Meentemeyer, 2006; Bartnicki and Williamson, 2012) [4] [14]-[17].

1.2. Wind Energy in Egypt

In fact, limited primary energy resources are consumed in Egypt. The most important of these resources are oil, natural gas, coal and hydropower. In addition, new and renewable energy resources such as solar and wind have a good potential. Due to its geographic location and varied topography, Egypt has some good locations with an average wind speed reaching 10 m/sec capable of producing competitive wind energy from wind turbines.

Since the early 1980's renewable energy has been considered as an integral part of the Egyptian policy. In a significant move, the Egyptian Government has launched a national program to apply new and renewable energy technologies. A new and Renewable Energy Authority (NREA) was established in 1986. The NREA's objectives are to introduce renewable energy technologies to Egypt on a commercial scale. The Egyptian Renewable Energy Development Organization (EREDO) was established in 1992 by mutual financing from Egypt and European Communities, covering renewable energy technologies, testing and endorsing certificates of components. The government program includes wind to supply energy to remote areas, which are located far from the grids, or directly to the grid in case of large wind power generation.

Recent achievements have been realized in the areas of institutional building, field testing, technology adoption and development of local wind industries. Currently, a wind farm of 6 MW is operative at Zafarana on the Gulf of Suez coast, where the first large commercial wind farm of 60 MW is being developed. Rapid penetration of different applications in many sectors is expected in the next few years, especially for the extension of the wind-electricity generation.

1.3. Description of the Red Sea Governorate

The Red Sea Governorate extends in a northwesterly direction between latitude 22°N 30°N. It is bordered from the east by the Red Sea and the Gulf of Suez coastal line and from the west by the eastern desert (**Figure 1**). The Red Sea coast line extends for 1080 km from the Gulf of Suez at latitude 29°N to the borders of Sudan at latitude 22°N. Area is equivalent to 120 thousand sq. km. and comprising six cities; Ras Gharib, Hurghada, Safaga, El Quseir, Marsa Alalm, and Shalateen. Between the Eastern Desert mountains and the Red Sea shoreline, there is a coastal plain of almost level land that extends along the edges of the Gulf of Suez and the Red Sea. The width of such ranges from 8 to 35 km.

The Red Sea Coast is characterized by a dominantly hot, dry and windy climate. In Hurghada city for example the annual average temperature is 24.5°C, the relative humidity is 79.6%, evaporation is 9.8 mm per day. The annual amount of rainfall is 4 mm. Thus the area is arid with sparse vegetation. The wind speed reaches 7.6 m/sec and of dominating NW direction. In some zones, the wind speed may reach 10 m/sec which is a high potential for wind energy.

The population of the Red Sea Governorate live in a number of cities on the coastline and few scattered villages in between. Tourist resorts are spreading down all coastlines. Egyptian Red Sea coast in general has very limited freshwater resources due to its geographical location in the arid sub-tropical region. The Egyptian Red Sea is also a major transportation route due to the Suez Canal and a key location for petroleum and natural gas production. The Red Sea Governorate is the richest province of the republic in mining and mineral potentials. It has the vast majority of metallic and non-metallic ores and ornamental stones and is quite rich in investment opportunities.

The economic base is partially tourism, which is nature based, occupying a large portion along the Red Sea shores. The Red Sea Governorate is rich in biodiversity providing potentials for ecotourism and nature preservation. Based on Shobrak *et al.* (2003) [17] and Baha El Din *et al.* (2003) [18], Egypt comprises a wide range of habitats critical for birds. These include: wetlands, high altitude mountains, desert wadis, coastal plains and marine islands (Egyptian Environmental Affairs Agency, 2014) [19]. The Red Sea islands of Tiran, Ashrafi, North Qeisum, Tawila and Zabargad are amongst the most important islands in the Egyptian Red Sea for breeding seabirds. About 150 of birds species are considered as resident with breeding populations (Baha El Din, 1999) [20]. Within birds populations, 16 species of global conservation concern have been recorded in Egypt (Collar *et al.*, 1994) [21]. The region is internationally famous for diving activities. Fishing is another important economic activity in the region (Baha El Din *et al.*, 2003) [18].

Development in Red Sea Governorate faces two main challenges; water and energy. The governorate has potentials for renewable energy such as solar and wind energy. Renewable energy can therefore be the backbone for development. Provision of necessary power can provide water by desalination of water from the Red Sea. Less than 1% of Egypt's current energy mix comes from wind, despite an abundance of wind resources, particularly in

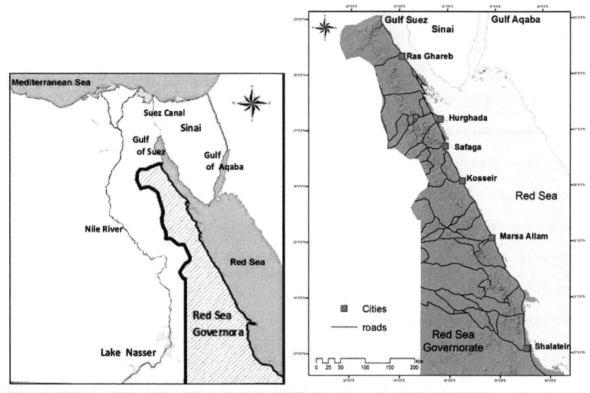

Figure 1. Location of the study area.

the Gulf of Suez area. As of 2008/2009, Egypt's Renewable Energy Authority (NREA), affiliated with the Ministry of Electricity manages Egypt's clean energy portfolio. It has installed 425 MW of wind power, including a wind farm at Zafarana, in Suez Governorate. Such wind farm has been operational since 2004 and has a capacity of 360 MW, where wind averages 9 meters/second (Shata and Hanitsch, 2006) [22]. A record summary of wind observations at some meteorological stations in Red Sea and Gulf of Suez is shown in **Table 1**.

2. Materials Used in the Study

Several data and information were collected, from the different sources to be used in order to achieve the overall objectives of this study, these include:

2.1. Remote Sensing Data: Optical and Radar Data Were Used for This Study as Follows

1) Landsat ETM+ images, to delineate shoreline and to investigate land use/land cover patterns.
2) Shuttle Radar Topography Mission (SRTM) elevation and derived land slope.

2.2. Maps

Various types of maps covering the study area have been compiled, digitized. All data were projected to WGS-84 of the Universal Transverse Mercator System (UTM) of geographic coordinates. The digitized layers were imported to a geographic database using ESRI ArcGIS 9.3 software [24].

These include:
1) Topographic maps at scale 1:50,000 published by the Egyptian General Survey Authority (1989) [25].
2) Climatic Atlas of Egypt, published by the Egyptian Meteorolgical Authority (1996) [26].
3) Wind Atlas of Egypt, published by Mortensen *et al.* (2005) [23].
4) Map of natural protectorates and birds sites, published by the Egyptian Environmental Affairs Agency (EEAA) [19].

A description of the main GIS layers used and sources is presented in **Table 2**.

3. Methodology

3.1. Theory and Approach

Spatial Multi-Criteria Evaluation (MCE) was conducted using analytical approach to determine suitable locations for the project. Analytical Hierarchy Process (AHP) is a multi-criteria decision method that uses hierarchical structures to represent a problem and then develop priorities for alternatives based on the judgment of the user (Saaty, 1980) [27]. The AHP procedure involves five essential steps (Lee *et al.*, 2008) [28]: 1) Develop the AHP hierarchy; 2) Pairwise comparison; 3) Estimate the relative weights; 4) Check the consistency; 5) Obtain the overall rating.

1) Develop the AHP hierarchy: In this step the complex problem is decomposed into a hierarchical structure with decision elements *i.e.*: objective, attributes or criterion map layers and alternatives (Vahidniaa *et al.*, 2008) [29].

2) Pairwise Comparison: AHP method is based on pair-wise comparison within a reciprocal matrix, in which the number of rows and columns is defined by the number of criteria. Accordingly, it is necessary to establish a comparison matrix between pairs of criteria, contrasting the importance of each pair with all the others. The comparison ratings are provided on a nine-point continuous scale, which was proposed by Eastman (1995) [30]. If we call that weight a_{ij}, and use that scale of comparison and if the relative weighting is $a_{23} = 3/1$, then the relative importance of attribute 3 with regard to 2 is its reciprocal $a_{32} = 1/3$. This process generates an auxiliary matrix in which the value in each cell is the result of the division of each value judgment (a_{ij}) by the sum of the corresponding column. Finally, the average of the normalized values of rows is obtained, which corresponds to the priority vector (W_j). This is normalized by dividing each vector value by n (the number of vectors), thus obtaining

Table 1. A record summary of wind observations at some meteorological stations in the Red Sea and Gulf of Suez: Data-collecting period, height above ground level of anemometer, data recovery rate (R), Weibull *A* and *K* parameters, mean wind speed (U), mean power density (E), and direction (DU), magnitude (U) of the mean wind vector. (Source: Mortensen *et al.*, 2005) [23].

Region/Station	period	Height [m]	R [%]	A [m·s⁻¹]	k	U [m·s⁻¹]	E [W·m⁻²]	D_U [deg]	U [m·s⁻¹]
Gulf Suez									
Ras Ghareb	5y	24.5	85.5	11.0	3.40	9.9	775	322	8.7
Red Sea									
Hurghada WETC	11y	24.5	79.6	7.6	2.32	6.7	308	322	4.9
Hurghada (62463)	10y	10.0	n/a	7.6	2.66	6.7	285	325	5.4
Kosseir (62465)	4y	10.0	97.1	5.1	2.03	4.6	178	334	3.5
Kosseir	4y	24.5	88.7	6.5	2.32	5.8	197	321	4.3

Table 2. Description of GIS data layers used in the current study.

GIS data layer	Description	Data Source
Vector (polygon)	Wind speed	Ministry of Electricity and Energy
Raster	Elevation	Shuttle Radar Topography Mission (SRTM)
Raster	Slope	Shuttle Radar Topography Mission (SRTM)
Vector (line)	Power lines	Topographic map, scale 1:50,000, Egyptian General Survey Authority 1989
Vector (line)	Highways	Ministry of Transportation
Vector (point)	Airport areas	General Organization for Physical Planning
Vector (polygon)	Birds sites	General Organization for Physical Planning
Vector (polygon)	National Protectorates	Egyptian Environmental Affairs Agency (EEAA)
Vector (point)	Cities and urban settlements	General Organization for Physical Planning
Vector (point)	Archaeology sites	General Organization for Physical Planning
Raster	Land use/land cover	Classification of Landsat ETM + imagery and FAO Africover data for Egypt
Vector	Shoreline	Digitized from Landsat ETM + imagery

the normalized overall priority vector, representing all factor weights (W_j) (**Table 3(a)**).

3) Estimate the relative weights: To determine the weighted sum vector we multiplying matrix of comparisons on the right by the vector of priorities to get a new column vector. Then divide first component of new column vector by the first component of priorities vector, the second component of new column vector by the second component of priorities vector, and so on. Finally, sum these values over the rows.

4) Estimation of the consistency vector: To calculate the consistency vector we divide the weighted sum vector by the criterion weights. Once the consistency vector is calculated it is required to compute values for two more terms, lambda (λ) and the consistency index (CI). The value for lambda is simply the average value of the consistency vector. The calculation of CI is based on the observation that λ is always greater than or equal to the number of criteria under consideration (n) for positive, reciprocal matrices and $\lambda = n$, if the pair wise comparison matrix is a consistent matrix. Accordingly, $\lambda - n$ can be considered as a measure of the degree of inconsistency and

$$CI = (\lambda - n)/(n - 1) \tag{1}$$

The term CI, referred to as consistency index, provides a measure of departure from consistency. To determine the goodness of CI. The consistency ratio (CR) can be defined as follows:

$$CR = CI/RI \tag{2}$$

where Random Index (RI) is the CI of a randomly generated pairwise comparison matrix of order 1 to 10 obtained by approximating random indices using a sample size of 500 (Saaty, 1980) [27]. **Table 3(b)** shows the value of RI sorted by the order of matrix.

The consistency ratio (CR) is designed in such a way that if CR < 0.10, the ratio indicates a reasonable level of consistency in the pairwise comparisons; if, however, CR > 0.10, then the values of the ratio are indicative of inconsistent judgments.

5) Overall rating: A weighted linear combination (WLC), a simple additive weight method was used to combine the criteria and constraints to yield the suitability map as follows:

$$S = \sum (W_j \times X_{ij}) \prod (\text{constraint map}) \tag{3}$$

S = composite suitability score,
W_j = weights assigned to each factor j,
X_{ij} = attribute score i of factor j.

Table 3. (a) AHP weighting scale (Saaty, 1980); (b) Random index (RI) (Saaty, 1980) [27].

(a)

Intensity / Relative importance	Definition (i in regards to a_j)	Values / a_{ij}	Numbers / a_{ji}
1	Equal importance	1	1
2	Intermediate	2	1/2
3	Moderate importance	3	1/3
4	Intermediate	4	1/4
5	Strong importance	5	1/5
6	Intermediate	6	1/6
7	Very strong importance	7	1/7
8	Intermediate	8	1/8
9	Extreme importance	9	1/9

(b)

Order Matrix	1	2	3	4	5	6	7	8	9	10
RI	0.00	0.00	0.58	0.9	1.12	1.24	1.32	1.41	1.45	1.49

3.2. Application on the Red Sea Governorate

3.2.1. Evaluation Criteria for Zoning Potential Wind Farms

Evaluation criteria are grouped into factors and constraints. Factors are criteria that promote the development of a certain activity. Constraints are concerned with the zones that could be negatively affected by a specific activity. It is thus such zones that should be avoided and/or protected from change. In GIS analysis, the constraints issue can be solved by keeping a set-back zones or buffer zones. A buffer zone is decided by setting a threshold value. The distance from a facility can be either a factor or a constraint or both. For example getting closer to a main road can be an economic factor that minimize transportation costs. Yet, getting too close to such road could be non-aesthetical, cause accidents, noise, compete with other land-use. It is therefore, necessary to reach a threshold that limits the factor from the constraint for a wind farm siting decision

3.2.2. Criteria and Thresholds

The preferences of each group for geographical locations use a set of specified criteria. For example, the environmentalists group can select maps of the impact of wind farms, or maps of the distances of wind farms ecologically sensitive areas due to bird collisions. Wind energy resource is the most important criteria for siting a wind energy farm. While performing site selection, both the wind energy potential and environmental acceptability/fitness need to be considered. A location which does not have sufficient wind energy potential is not an appropriate location for wind turbines no matter how high it satisfies all acceptability objectives and concerns. A conceptual flow chart for the criteria and methodology is shown in (**Figure 2**).

3.2.3. Wind Power Density

According to (Mortensen *et al.*, 2005) [23] the wind resources of Egypt have recently been assessed by the New and Renewable Energy Authority, the Egyptian Meteorological Authority and Risø National Laboratory; (NREA) the results are reported in detail in a Wind Atlas for Egypt. The primary purpose of the Atlas is to provide reliable and accurate wind atlas data sets for evaluating the potential wind power output from large electricity-producing wind farms. Numerical wind atlas methodologies have been used to solve the issue of insufficient wind measurements based on meteorological stations. Based on Mortensen *et al.* (2005) [23], the KAMM/WAsP method was developed and used by Riso laboratory to produce estimates of wind resource on a national scale. A predicted map for wind climate of Egypt was determined by meso-scale modeling. The map shows the mean wind speed in [m·s^{-1}] at height of 50 m over the actual land surface. The horizontal grid point resolution is 7.5 km.

The wind power in a given site depends on having sufficient wind speed available at the height at which the turbine is to be installed (Vanek and Albright, 2008) [31]. Wind power density is a most important factor because it provides information on the most feasible and profitable areas in the region for siting a wind power project (Baban and Perry, 2001) [4]. Bartniki and Williamson, 2012 [15] explain that wind power density is a function of the area's average wind velocities and the air densities, which involves land elevations. The wind power density is derived by Equation (4) and Equation (5) (Hughes, 2000) [32].

$$\rho = 1.225 - \left[\left(1.194 \times 10^{-4} \right) \times \text{elevation} \left(m \right) \right] \tag{4}$$

$$wpd = 1/2\, \rho V^3 \tag{5}$$

where: V = average wind speed (m/s); ρ = air density (kg/m^3); wpd = wind power density (watt/m^2).

Shuttle Radar Topography Mission SRTM digital elevation model (figure) was used to model the air density using (Equation (4)). The wind speed energy map (V) developed by Mortensen *et al.*, 2005 [23] was downloaded geometrically corrected, digitized, projected , converted to raster format and a resample function was applied. The data was finally used to calculate the wind power density using Equation (5).

3.2.4. Power-Lines Proximity and Setback Buffer

A suitable location will need to be in close proximity to existing roads and hydro lines to minimize production costs (Nextra Energy Canada, 2011 [33]; Bartnicki and Williamson, 2012 [15]). The distance to transmission lines is a necessity in order to transport the energy created by the wind turbines and reduce costs. Land that is connected to an electrical grid therefore provides a more suitable site. For setback distance, according to Moiloa 2009 [34] and the DEADP, 2006 [35] suggests a minimum distance of 250 meters should be kept apart from high voltage

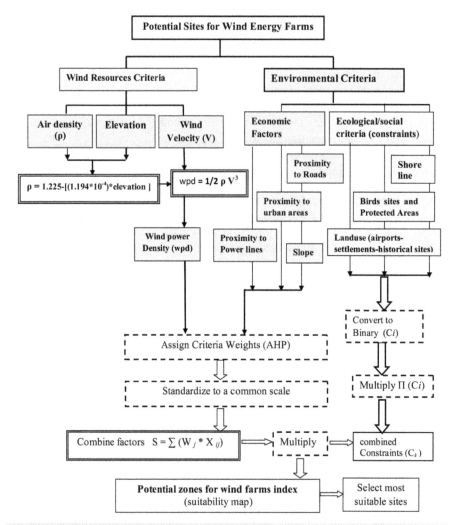

Figure 2. A conceptual chart for the applied methodology.

lines. Same distance was considered for the current study.

3.2.5. Roads Proximity and Setback

Wind farms should be set apart from roads and railroads. According to Moiloa, 2009 [34] and the DEAP 2006 [35] suggested a distance of 2.5 km from main roads and railroads and a distance of 500 meter from secondary roads. Construction costs start to increase the further away a project is from existing roads due to the need for new road construction (Bartniki and Williamson, 2012) [15]. A distance function was used to classify the region, where the areas closer to the roads setback buffers were given high suitability while areas further to the buffer were given low suitability. For the present study, a setback of 500 m was used.

3.2.6. Urban Areas and Cultural Sites Proximity and Setback

CNdV Africa 2006 [36] report defines heritage sites as sites of historical and cultural value either with national or provincial designations. These sites are considered cultural amenities that should be protected. Bartnicki and Willamson, 2012 [15] used a 550 m setback for urban, recreational and historic areas are granted. For the current study, a setback of 1000 m was used for historic sites while 2000 m was used for cities to account for growth expansion.

3.2.7. Slopes

Luo *et al.*, 2007 [37], Bartnicki and Williamson 2012 [15] explain that at the summit of steep slopes wind may not hit the turbine rotor at a perpendicular angle. This will result in an increased level of fatigue for the turbine. Thus a slope of greater than 5 degrees will yield more turbulent wind patterns causing disruptions in turbine stability. Building on higher slopes also increases project costs. Ideally, the terrain should be rounded or flat because they will be exposed to higher, more constant wind speeds (Baban and Parry, 2001) [4]. The slope was reclassified to three categories with most suitable for flat to 5 degrees and unsuitable for slopes greater than 10 degrees and marginally suitable for values in between.

3.3. Ecological and Social Factors (Exclusion Zones/Constraints)

Ecological and social evaluation criteria can be factors or constraints. A constraint or an exclusion zone is restrictedly unsuitable zone for wind turbine installation. It is excluded for protecting effects on environment, communities, visualization, eco-conservation, and engineering frontier (Bennui et al., 2007) [7]. For constraint criteria, a threshold was assigned. Such threshold classifies a criteria raster into suitable and unsuitable pixels using a binary classification (Effat and Hegazy, 2013; Effat, 2014) [38] [39]. Suitable pixels are assigned a value of "one" while unsuitable pixels are assigned a "zero" value. In the last stage of data analysis, inappropriate zones will be combined in a single constraints binary map and excluded. Constraints for the present study are explained as follows:

3.3.1. Shoreline Setback
Wind farms should be set apart from inhabited areas. According to Moiloa, 2009 [34] and the DEAP, 2006 [35] suggested a setback distance of 4 km from the coast. A similar buffer zone was used around the shoreline considering possible birds flight paths and future tourism marine activities.

3.3.2. Protected Areas (Cultural, Ecological and Birds Areas)
Regionally important geological/geomorphologic sites, and other natural reserves are protected by national legislations. Such lands were considered constraints as the development of a wind farm may have a significant impact on the environmental values of such areas. According to Moiloa, 2009 [34], the DEAP 2006 [35] a distance of 500 meter from nature reserves and a distance of 1 km from national protectorates, nesting areas and flying routes of protected species is appropriate. CNdV Africa 2006 [36] used a distance of 2 km from national protectorates for the development of Western Cape, South Africa. For the present study, a distance of 1 km around the natural protectorates and ecological sites was considered appropriate.

3.3.3. Land-Use—Land-Cover
Certain land-use have to be set apart from wind farms. Airports is one the most important as navigation can be affected by such operations. CNdV Africa (2006) [36] applied a distance of 25 km from airport with primary radar for the development of Western Cape. Same distance was adopted as buffers for the airports.

3.3.4. Developing a Pairwise Comparison Matrix
The Pairwise Comparison Matrix used for the present study was developed based on the review of relevant literature. However, there could be a different judgment for the relative magnitude of the criteria when these are compared in pairs. The decision making process in the multiple criteria problems is a subjective process depending on the decision maker vision.

3.4. Standardization of the Criteria

For each factor, the attributes were standardized by transforming the original values to a suitability value using Saaty's, (1980) [27] suitability scale from 1 - 9. The higher value is more favorable and vice versa. Areas deemed as constraints will have a suitability score of zero. Areas with a higher suitability will have a higher score (Bartnicki and Williamson, 2012) [15].

3.5. Combining the Criteria

Evaluation criteria were combined in a grid that contains all standardized grids calculated from each of the individual grids. All the input evaluation criteria were in standardized raster grid format and with a 100 m cell size. Equation (3) generates a suitability grid that ranks the suitable areas by percentage.

4. Results and Discussion

The study was conducted using satellite data from Landsat ETM+ and SRTM and geographic information systems (GIS) for mapping the suitability of the Red Sea Governorate for siting wind farms. Results revealed that coastal areas along the Gulf of Suez and the northern zones of the investigated region have high wind energy potentials. Such zones are appropriate for setting up electricity generating wind turbines. The total investigated area is equivalent to 120 thousand sq. km. Most suitable zones amounts to 706.5 sq. km with suitability percent ranging between 83% and 100% and highly suitable zones amounting to 3781 sq. km having suitability percent that range from 66% to 83%. The results reveal that the northern part of the investigated area is quite rich in wind power potentiality. Results can be explained as follows:

4.1. Wind Power Map (Wind Resources Criteria)

The study revealed that coastal areas along the Gulf of Suez and the northern zones of the Red Sea coast with an area equivalent to 120 thousand sq. km have high wind energy potentials. Such zones are appropriate for setting up electricity generating wind turbines. The calculations of wind power density from the wind speed maps reveal a maximum value is around 660 watt/m^2 and a mean value of 97 watt/m^2. The elevation, air density, average wind velocity and the resultant wind power density maps are presented in **Figures 3(a)-(d)** respectively.

4.2 Economic Criteria Maps and Standardization

The standardized economic criteria maps for distance to roads, distance to power lines, distance to settlements are depicted in **Figures 4(a)-(c)**. The slope suitability attrubyte map is shown in **Figure 5**. Assigned weights resulting from applying Analytical Hierarchy Process (AHP) are presented in **Table 4**. The standardized attribute values based on suitability for siting wind energy zones are presented in **Table 5**.

4.3. Ecological and Cultural Criteria Maps (Constraints)

The ecological and cultural values criteria maps are binary maps. Such maps are presented in **Figure 6**. In each of the constraints maps, the excluded zones are presented in black color while the rest of the study area is presented in white color. **Figures 6(a)-(f)** depict the excluded zones for the national protectorates, shoreline, airports, cities, main roads and birds sites respectively. The set back buffer zones are described in **Table 6**.

4.4. Suitability Map

The suitability index values resulting from the overlay (suitability function) was classified into five suitability classes. Each class representing a suitability range (**Table 7**).

Most suitable zones of class 5 have wind velocity that range between 8.0 to 11.0 m/sec. and wind power density ranging between 543 to 659 watt/m^2 is a huge energy potentiality and flat lands. Such class exists along the north-east shores of Gulf Suez and Red Sea and extend inlands towards the west. Two airports exist in such zone namely Ras Shukhayr new airport and Ras Jismah new airport. Such airports necessitates setback buffers each of which is assigned an area of 20 - 25 km for proper navigation needs. In addition, ecological areas for bird flight sites and high slope lands were excluded. Suitable net areas for class 5 is equivalent to 706.7 sq. km with suitability values from 84% -100% (**Figure 7**).

Highly suitable zones in class 4 area have wind power density ranging between 310 - 425 watt/m^2 and a land area equivalent to 3781.2 sq. km after subtraction of setback buffers and constraints. This class exists almost parallel to the Gulf Suez and the northern parts of the Red Sea , mostly flat lands to very gentle slopes (less than 9 degrees) and are in close proximity to class 5 (**Figure 7**). Suitability values for this class range from 65% to 83%.

Moderately suitable zones in class 3 have wind velocity range of 4.5 - 6.4 m/sec which is acceptable to good, but the slopes are quite steep. The slope constraint cause more turbulent wind patterns and may cause disruptions in turbine stability and also reduces the suitability values. Net area of such class is around 35,300 sq. km with suitability values ranging from 50% to 65% (**Figure 7**).

Less suitable zones in class 2 have wind velocity range of 4.5 - 5.0 m/sec. which is marginally acceptable. Steep slopes exceed the accepted threshold (greater than 9 degrees) in such areas which in turn lead to low sui-

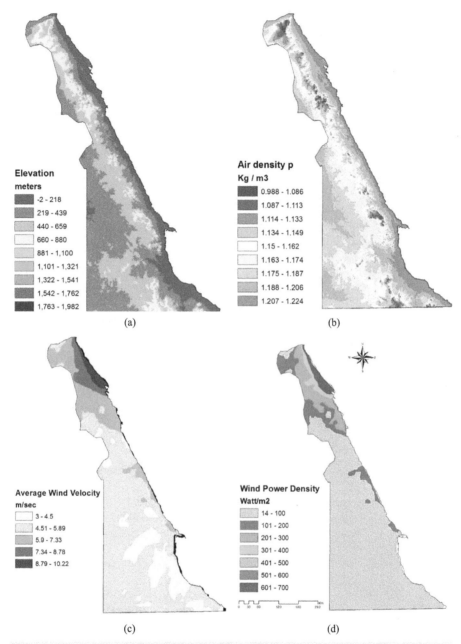

Figure 3. Wind power density evaluation criteria (grids in true value categories) (a) elevation; (b) air density; (c) average wind velocity; and (d) wind power.

tability values. Total area of such class is equivalent to 45,690.7 sq. km with suitability values ranging from 31% - 49% (**Figure 7**).

　　Unsuitable zones in class 1 are excluded zones (zones treated as constraints). Total area of such class is equivalent to 35,298.8 sq. km and a suitability range (0 - 30%). Regardless of the wind velocity, such zones were excluded for ecological, social and economic reasons (**Figure 7**).

5. Conclusion and Recommendation

The current study employed remotely sensed data in a GIS-based multi-criteria evaluation model to identify and map appropriate zones for wind energy farms. The method was quite flexible in creation of the evaluation criteria, assigning importance weights, standardization and map overlay. It provides visual intermediate and final results in

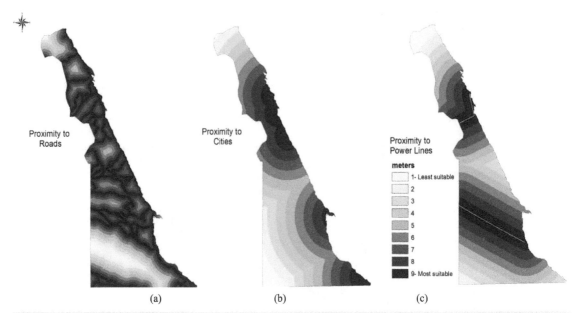

Figure 4. Standardized suitability grids for economic evaluation criteria: (a) proximity to roads; (b) proximity to cities; (c) proximity to power lines.

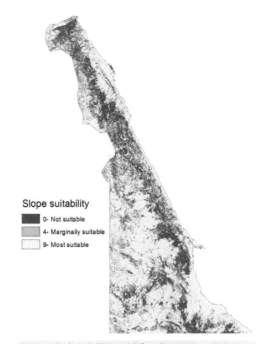

Figure 5. Suitability grid for slope.

Table 4. Pairwise comparison matrix for criteria weight assignment.

Criteria	Wind power density	Slope (degrees)	Proximity to power lines	Proximity to roads	Proximity to urban centers	Calculated weight
Wind power density (W/m²)	1	4	7	9	9	0.618
Slopes	1/4	1	7/4	9/4	9/4	0.155
Proximity to Power Lines	1/7	4/7	1	9/7	9/7	0.088
Proximity to Roads	1/9	4/9	7/9	1	1	0.068
Proximity to Urban centers	1/9	4/9	7/9	1	1	0.068

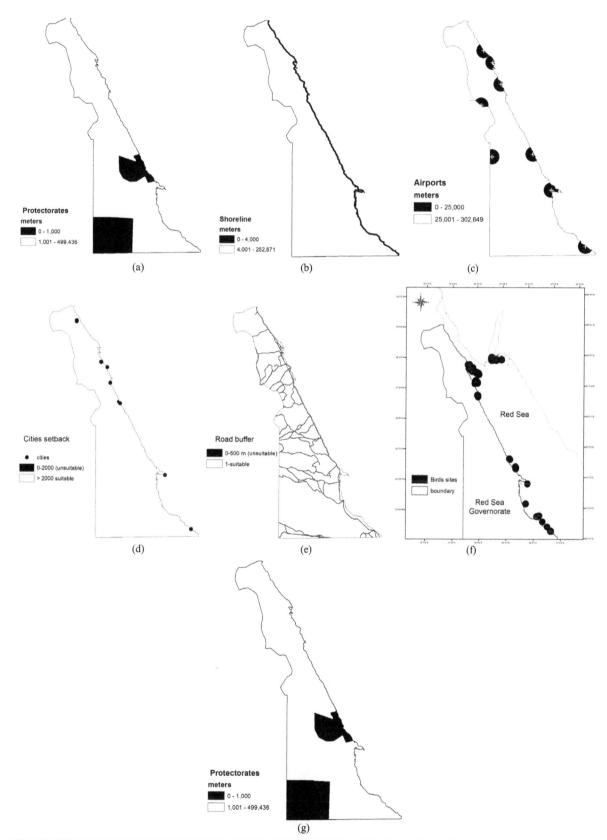

Figure 6. Ecological and cultural (constraints): (a) natural Protectorates; (b) shoreline; (c) airports; (d) cities and historical sites; (e) roads; (f) birds protection sites; (g) combined constraints map.

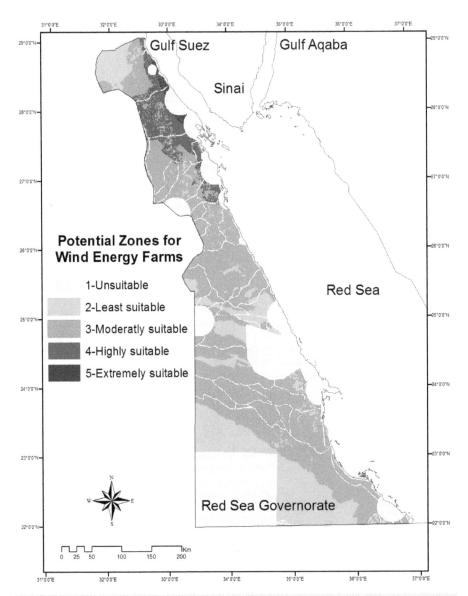

Figure 7. Suitability index for most appropriate zoning of wind farms in Red Sea Governorate.

Table 5. Standardization of the wind farm site evaluation criteria.

Standardization of the evaluation criteria standardization of evaluation factors					
Suitability scale	Wind power density (watt/m²)	Slope degrees	Proximity to power lines (m)	Proximity to cities (m)	Proximity to roads (m)
0		>=9	0 - 249	0 - 2000	0 - 500
1	14 - 86		214,445 - 265,452	286,108 - 321,870	68,713 - 77,301
2	87 - 158		176,969 - 214,444	250,344 - 286,107	60,124 - 68,712
3	159 -229		147,822 - 176,968	214,581 - 250,343	51,353 - 60,123
4	230 -301	>5 < 9	123,879 - 147,821	178,818 - 214,580	42,946 - 51,352
5	302 - 373		100,977 - 123,878	143,054 - 178,817	34,356 - 42,945
6	374 - 445		77,034 - 100,976	107,291 - 143,053	25,768 - 34,356
7	446 -516		52,050 - 77,033	71,528 - 107,290	17,179 - 25,767
8	517 -588		24,985 - 52,049	35,764 - 71,527	8590 - 17,178
9	589 - 660	<=5	250 - 24,984	2000 - 35,763	500 - 8589

Table 6. Excluded zones (Constraints) threshold.

Criteria (excluded zones)	Buffer zone (m) around excluded zones
Protectorates	1000 m around protectorates
Shoreline	4000 m around shoreline
Birds sites	1000 m around birds sites
Power lines	250 m around power lines
Cities	2000 m around cities
Historic sites	1000 m around historic sites
Roads	500 m around roads
Airport	25,000 m around main airports
Slopes	More than 9 degrees

Table 7. Suitable classes and land areas for wind power energy crop.

Class	Rang and percentage of suitability values	Class description	Area (sq. km)	Percentage of total area
1	0 - 0.30	Unsuitable	35298.8	0.294
2	0.31 - 0.49	Least suitable	45690.7	0.380
3	0.50 - 0.65	Suitable	35300.7	0.294
4	0.66 - 0.83	Highly suitable	3781.2	0.031
5	0.84 - 1.00	Extremely suitable (ideal)	706.5	0.005

the form of thematic maps that are comprehensive and useful for planning purposes. The method succeeds in mapping potential zones rich in wind power energy and avoiding sensitive ecological sites, while considering some economic factors such as slopes, accessibility and power network. By excluding the land constraints, the developed model identified the ideal zones with all suitability conditions for siting wind energy farms in the northern Red Sea Coast. These ideal zones amount to 706 km·sq. with suitability values of 83% - 100%. Also, highly suitable zones amounting to 3781 km·sq. with suitability values of 66% - 83% have been determined. Thus multi-objectives could be reached. The results of the current study highlights the need of adopting a holistic integrated approach that integrates land resources, potentials and constraints in the land-use decision strategies for achieving sustainable planning at a regional scale. Applying such techniques unveils the relationship between potential resources and vulnerable features that can be spatially competing. Thus providing indicator maps; as a guide for zoning and land-use strategies. The methodology is applicable and can be conducted for creating zoning maps of wind power energy in other regions.

Based on the output of this study, potential zones in each individual location should be field investigated and an environmental impact assessment to be further assessed. It is highly recommended that the land-use decision makers adopt the Spatial Multicriteria Evaluation analysis. Such method is believed to overcome the shortcomings of the current planning practice especially in developing countries, due to need of a multidisciplinary approach and evaluation tools in the land-use planning decisions.

References

[1] Fernandez, L.M., Saenz, J.R. and Jurado, F. (2006) Dynamic Models of Wind Farms with Fixed Speed Wind Turbines. *Renewable Energy*, **31**, 1203-1230. http://dx.doi.org/10.1016/j.renene.2005.06.011

[2] Tester, J.W., Drake, E.M., Driscoll, M.J., Golay, M.W. and Peters, W.A. (2005) Sustainable Energy; Choosing among Options. The MIT Press, Cambridge.

[3] Ramirez-Rosado, I.J.R., Garrido, E.G., Jimerenz, L.A.F., Zorzano-Santamar, P.J., Monteiro, C. and Miranda, V. (2008) Promotion of New Wind Farms Based on a Decision Support System. *Renewable Energy*, **33**, 558-566. http://dx.doi.org/10.1016/j.renene.2007.03.028

[4] Baban, S.M.J. and Parry, T. (2001) Developing and Applying a GIS-Assisted Approach to Locating Wind Farms in the UK. *Renewable Energy*, **24**, 59-71. http://dx.doi.org/10.1016/S0960-1481(00)00169-5

[5] Nguyen, K.Q. (2007) Wind Energy in Vietnam: Resource Assessment, Development Status and Future Implications. *Energy Policy*, **35**, 1405-1413. http://dx.doi.org/10.1016/j.enpol.2006.04.011

[6] Yue, C.D. and Wang, S.S. (2006) GIS-Based Evaluation of Multifarious Local Renewable Energy Sources: A Case

Study of the Chigu Area of Southwestern Taiwan. *Energy Policy*, **34**, 730-742.
http://dx.doi.org/10.1016/j.enpol.2004.07.003

[7]   Bennui, A., Rattanamanee, P., Puetpaiboon, U., Phukpattaranont, P. and Chetpattananondh, K. (2007) Site Selection for Large Wind Turbines Using GIS. *PSU-UNS International Conference on Engineering and Environment-ICEE-2007*, Phuket, May 10-11, 2007, 90112.

[8]   Patel, B. and Rosier, A. (2013) Basic Criteria for Wind Project Site Selection and Optimization. ECOWAS Regional Workshop on Wind Energy. Pria, Cape Verde.
http://www.ecreee.org/sites/default/files/event-att/wind_project_site_optimization.pdf

[9]   Elliasson, B. (1998) Energy and Global Changes. ABB Corporate Research.

[10]   Walker, J.F. and Jenkens, N. (1997) Wind Energy Technology. John Wiley & Sons, Chichester.

[11]   Afgan, N.H. and Carvalho, M.G. (2002) Multi-Criteria Assessment of New and Renewable Energy Power Plants. *Energy*, **27**, 739-755. http://dx.doi.org/10.1016/S0360-5442(02)00019-1

[12]   Malczewski, J. (1999) GIS and Multicriteria Decision Analysis. John Wiley and Sons, Inc., New York.

[13]   Wood, L.J. and Dragicevic, S. (2007) GIS-Based Multicriteria Evaluation and Fuzzy Sets to Identify Priority Sites for Marine Protection. *Biodiversity and Conservation*, **16**, 2539-2558. http://dx.doi.org/10.1007/s10531-006-9035-8

[14]   Mann, D., Lant, C. and Schoof, J. (2012) Using Map Algebra to Explain and Project Spatial Patterns of Wind Energy Development in Iowa. *Applied Geography*, **34**, 219-229. http://dx.doi.org/10.1016/j.apgeog.2011.11.008

[15]   Bartnicki, N. and Willamson, M. (2012) An Integrated GIS Approach to Wind Power Site Selection in Huron County, Ontario. Department of Geography, University of Guelph, Guelph.

[16]   Rodman, L.C. and Meentemeyer, R.K. (2006) A Geographic Analysis of Wind Turbine Placement in Northern California. *Energy Policy*, **34**, 2137-2149. http://dx.doi.org/10.1016/j.enpol.2005.03.004

[17]   Shobrak, M., Al-Suhaibany, A. and Al-Sagheir, O. (2003) Regional Status of Breeding Seabirds in the Red Sea and the Gulf of Aden. The Regional Organization for the Conservation of the Environment of the Red Sea and Gulf of Aden (PERSGA). http://www.persga.org/Files/Common/Sea_Birds/Reginal_Status_of_Seabirds.pdf

[18]   El-Din, M.B., El-Din, S.B. and Shobrak, M. (2003) Status of Breeding Seabirds in the Egyptian Red Sea. Report for PERSGA, Jeddah, 30 p.

[19]   Ministry of State of Environmental Affairs, Egyptian Environmental Affairs Agency (EEAA) (2014) Important Bird Areas (IBAs) of Egypt. http://www.eeaa.gov.eg/english/main/protect_bird.asp

[20]   El-Din, S.B. (1999) Directory of Important Bird Areas in Egypt. Birdlife International, The Palm Press, Cairo.

[21]   Collar, N.J., Crosby, M.J. and Stattersfield, A.J. (1994) Birds to Watch 2. Birdlife International, Cambridge.

[22]   Shata, A.S. and Hanitsch, R. (2006) The Potential of Electricity Generation on the East Coast of Red Sea in Egypt. *Renewable Energy*, **31**, 1597-1615. http://dx.doi.org/10.1016/j.renene.2005.09.026

[23]   Mortensen, N.G., Hansen, J.C., Badger, J., Jørgensen, B.H., Hasager, C.B., Youssef, L.G., Said, U.S., *et al.* (2005) Wind Atlas for Egypt, Measurements and Modeling. Wind Atlas for Egypt: Measurements, Micro and Mesoscale Modeling. http://www.windatlas.dk/egypt/download/wind%20atlas%20for%20egypt%20paper%20(menarec3).pdf

[24]   (2006) ESRI Arc Map Help. Version 9.2, User Manual, ESRI, Redlands.

[25]   The Egyptian General Survey Authority (1989) Topographic Maps at Scale 1:50,000.

[26]   Egyptian Meteorological Authority (1996) Climatic Atlas of Egypt.

[27]   Saaty, T.L. (1980) The Analytic Hierarchy Process. McGraw-Hill, New York, 20-25.

[28]   Lee, A.H.I., Chen, W.C. and Chang, C.J. (2008) A fuzzy AHP and BSC Approach for Evaluating Performance of IT Department in the Manufacturing Industry in Taiwan. *Expert Systems with Applications*, **34**, 96-107.
http://dx.doi.org/10.1016/j.eswa.2006.08.022

[29]   Vahidnia, M.H., Alesheikhb, A., Alimohammadic, A. and Bassiri, A. (2008) Fuzzy Analytical Hierarchy Process in GIS Application. *The International Archives of the Photogrammetry, Remote Sensing and Spatial Information Sciences*, **37**, 593-596.

[30]   Eastman, J.R., Jin, W., Kyem, P.A.K. and Toledano, J. (1995) Raster Procedures for Muli-Criteria/Multiobjective Decisions. *Photogrammetric Engineering & Remote Sensing*, **61**, 539-547.

[31]   Vanek, F.M. and Albright, L.D. (2008) Energy Systems Engineering-Evaluation and Implementation. McGraw-Hill, New York.

[32]   Hughes, T. (2000) Calculation of Wind Energy and Power. Lesson Number 1 in an Oklahoma Wind Power Tutorial Series. http://www.seic.okstate.edu/owpi_old/about/library/lesson1_windenergycalc.pdf

[33]   Nextra Energy Canada (2011) Conestogo Wind Farm, Revised Construction Plan Report. Nextra Enegy Canada,

Conestogo Project.
http://www.nexteraenergycanada.com/pdf/conestogo/Revised_Conestogo_Construction_Report_FINALv1.pdf

[34] Moiloa, B.H.E. (2009) Geographical Information Systems for Strategic Wind Energy Site Selection. M.Sc. Thesis, Faculty of Earth and Life Sciences, Vrije University, Amsterdam.

[35] DEADP (2006) Strategic Initiative to Introduce Commercial Land Based Wind Energy Development to the Western Cape: Towards a Regional Methodology for Wind Energy Site Selection, Report Series 1-7. Department of Environmental Affairs and Development Planning, Cape Town.

[36] CNdV Africa Planning and Design (2006) Strategic Initiative to Introduce Commercial Land Based Wind. Energy Development to the Western Cape. Towards a Regional Methodology to Wind Energy Site Selection. Report 2. Methodology 1. Prepared for Provincial Government Wind Energy Landscapes: Specialist Study: Report 2. Methodology 1, 9-11. http://ebookbrowsee.net/3-report-2-method-1-criteria-based-assessment-pdf-d251378713

[37] Luo, C., Banakar, H., Shen, B. and Ooi, B.T. (2007) Strategies to Smooth Wind Power Fluctuations of Wind Turbine Generator. *IEEE Transactions on Energy Conversion*, **22**, 341-349. http://dx.doi.org/10.1109/TEC.2007.895401

[38] Effat, H.A. and Hegazy, M.N. (2013) A Multidisciplinary Approach to Mapping Potential Urban Development Zones in Sinai Peninsula, Egypt, using Remote Sensing and GIS. *Journal of Geographic Information System*, **5**, 567-583 http://dx.doi.org/10.4236/jgis.2013.56054

[39] Effat, H.A. (2014) Resource-Based Zoning Map for Sustainable Industrial Development in North Sinai using Remote Sensing and Multicriteria Evaluation. *International Journal of Sustainable Development and Planning*, **9**, 119-134.

3

Landslide Hazard Mapping of Nagadhunga-Naubise Section of the Tribhuvan Highway in Nepal with GIS Application

Arjun Raj Pandey[1], Farzad Shahbodaghlou[2]

[1]Department of Civil, Architectural, and Environmental, Illinois Institute of Technology, Chicago, USA
[2]Departmant of Engineering, California State University, East Bay, Hayward, USA
Email: apandey2@hawk.iit.edu, farzad.shahbodaghlou@csueastbay.edu

Abstract

The aim of this project was to prepare and study a hazard map of Nagadhunga-Naubise section of the Tribhuvan highway. This section lies in the Middle Mountain region of Nepal. For the preparation of the hazard map of the corridor three steps, initial study, field investigation, and data analysis and presentation were carried out. In the initial study, the collection of available data and review of the literature were done. The base map was then prepared from the topographical map. In the field investigation step, all information and maps prepared earlier in the initial study were verified by field check. In the final step, prepared and verified data were then analyzed for the hazard mapping. Topography (gradient, slope shape and slope aspect), geology, drainage and landuse were considered to be the major influencing factors in the slope stability. Pre-assigned hazard rating method was used for hazard mapping of the study area. The area was divided into equal facets. Then ratings of responsible factors to the hazard were assigned to each facet and overlaid based upon a predetermined rating scheme. Total estimated hazard was the sum of these ratings for each overlay. Hazard map was prepared by using three categories as low hazard, medium hazard and high hazard. The Geographic Information System (GIS) was the main tool for the data input, analysis, and preparing of the final hazard map. The hazard map showed the areas of different hazard potential classes of; "low" with 32% portion, "Medium" with 51%, and "high" with 17% portion.

Keywords

Geology, Topography, Hazard Map, Hazard Rating, GIS (Geographic Information System)

1. Introduction

Nepal is a mountainous country of geologically young mountains [4] [5]. Generally roads and other linear structures are laid along the steep hill slopes. These hill roads are also crossed by numbers of streams and rivers. Tribhuvan highway is also one of the major roads laid on the steep hill slopes, which joins the Capital City to the other parts of the country. Roads laid on such fragile geomorphology are vulnerable to various kinds of disasters [5] [10]. The roads section Nagadhunga-Naubise also experiences frequent slope failures, landslide, and roadblock [6] [7]. The landslide and blockage of the road especially during monsoon season have frequently blocked the traffic movement with numbers of accidents [7]. Landslides have always been a major hazard to the safety of people and their property. The fragile and young Himalayas with torrential monsoon, earth tremors and environmental degradation have together increased occurrence of landslide in Nepal.

Increasing population density and expansion of infrastructure have resulted in changes in natural slope, river morphology, and land-use/land cover. The frequency in the occurrence of the landslides, and flood hazards have been increasing in recent years [2] [5] [6] [10]. This is evidenced by the fact that in Nepal, floods and landslides alone claimed more than 5800 human lives during the period of 1983 to 2003 [8]. The loss of lives, owing to floods, landslides, and avalanches, comprises about 29% of the total losses from all types of disasters. As a result of increased natural hazards, the magnitudes of lives and property at risk have also increased correspondingly [11] [12].

In this study the Nagadhunga-Naubise road corridor (85°00' to 85°12.50' longitude and 27°38' to 27°48' latitude), six kilometer square area, 500 m on either side of the highway section was selected as the study area.

Hence, the identification of hazards/susceptible areas and the concurrent assessment of the risks associated with these hazards become important for devising effective mitigation plans. In this context an attempt has been made to prepare a landslide hazard map of the area.

The preliminary examination of the area showed that 11 landslides of different types and sizes were identified [8]. Out of these, 8 landslides were investigated in more detail during fieldwork. The selection of these slope failures depended upon its severity and impact on infrastructures. The detail investigation of the landslides in the area is to understand the triggering factors of landslide in the area, which ultimately became the basis for the hazard rating. This gave a strong basis for the engineering judgment of the rating of hazards as well.

2. Research Objectives

The objective of the study is to prepare a hazard map of the section Nagadhunga-Naubise (500 m either side) road section. The detailed studies of the individual landslide were conducted to understand the landslide triggering factors in order to develop of the system of hazard rating. The study is primarily based on available maps (topographical and geological), aerial photos, visual interpretation and relevant digital imageries and other secondary sources of information. In addition, detailed field surveys were carried out to verify and update the information obtained from maps, aerial photos, and imageries. The Geographic Information System (GIS) technology is the main tool used for data input, analysis, and hazard mapping. Four parameter maps: Topography (gradient, slope shape, slope aspect), Land-use, Geology, and Drainage, were used for landslide hazard mapping [6].

3. Research Approaches

This study is done in three parts: Defining a research hypothesis and relevant research questions, conducting a literature review based on the research design, survey and data collection from field and other sources, and analyzing and presenting the research.

There have been several methodologies developed for the Hazard Mapping [1] [5] [6]. These methods can be broadly categorized into three types: Simple qualitative mapping (pre-assigned rating method); Statistical (bivariate and multivariate) method; and Deterministic method. The Statistical and Deterministic method are particularly applied for landslide hazard mapping. The Statistical method is used to prepare the Hazard Map in this study. Arc View is used to produce Digital Elevation Model (DEM) from which several landslide hazard parameters such as slope gradient, slope shape, relative relief, and the drainage density were derived in raster/grid format.

3.1. Development of System for Hazard Mapping

For this study four categories of factors, which contribute landslide, were considered: topography, geology, land use, and river/drainage system. In addition, the category of topography was divided into four sub-groups: respective gradient, vertical profile, horizontal profile, and aspect.

As stated earlier, landslide hazard is governed by the combination of various factors, which have roles of differing importance [6] [11]. Based on relative importance of each factor a numerical weighting (**Table 1** column 2) were assigned. The same was done for the sub-groups of topography and geology (**Table 1** column 4). The selection of these factors and weights are based on engineering judgment, available statistical data and findings from other studies.

Table 1. Quantification of hazard zonation (Deoja and *et al.*, 1991 [5]; Moser, 2002 [9]; Sarkar and *et al.*, 1995 [10]).

1	2	3	4	5	6	7
Category of Factors	Weight among Categories	Sub-Group	Weight Of Sub Group	Global Weight	Criteria	Failure Potential
Topography	30	Gradient	0.4	12.0	0 - 20	0.00
					20 - 25	0.25
					25 - 30	0.50
					30 - 40	0.75
					40 - 90	1.00
		Shape of Vertical Profile	0.4	12.0	Concave	0.00
					Straight	0.50
					Convex	1.00
		Shape of Horizontal Profile	0.1	3.0	Concave	0.00
					Straight	0.50
					Convex	1.00
		Slope Aspect	0.1	3.0	SE, S	0.00
					E, W, SW	0.50
					NE, NW, N	1.00
Land Use	20	Land Use Type	1.0	20.0	Thickly Vegetated	0.00
					Forest	0.50
					Cultivated Bari (Sloping Terrace)	0.60
Geology	25	Lithology	1.0	25	Colluviums	0.00
					Sopayang Formation (Dark Argillaceous)	0.60
					Chandragiri Limestone	0.5
					Weak Rocks	0.7
River/Drainage System	25		1.0	25	0 - 15 m	0.00
					15 - 30 m	0.25
					>30 m	0.50

To obtain a comparable global weight (column 5) of each factor, the values of column 2 and 4 were multiplied. **Table 1** assigns to the categories of land use, geology and drainage followed by gradient and shape of vertical profile with a global weight value of 12. Aspect and horizontal profile have lowest importance; hence, their values of 3.

It should be noted that **Table 1** is the tabular form of a two-level decision tree adopted fromre-evaluation of failure potential.

To evaluate the hazard potential of a single cell, a criterion of each sub group was established (**Table 1**, column 6). These criteria were rated according to third slope failure potential which is listed in column 7. The values range from 0 to 1, where the higher the values the higher are the potential slope failure.

3.2. Procedure for Mapping of Landslide Hazard

The area in which the mapping was conducted had to be divided into a number of cells. These cells are represented by squares of an area of 5 m by 5 m. For each cell, the slope failure potential was assessed using the established criteria shown in the **Table 1**. To obtain the slope failure potential map, data layers for each factor have been produced and superimposed. The development of these maps is described in the following sections. The **Figure 1** is the schematic illustration of superimposition of different layers to produce the hazard map.

4. Findings and Discussions

The study area is divided into a number of 5 m by 5 m cells. The total area covering the section is about 6 km². Each cell of the area has been rated for its landslide hazard potential based on the criteria in the **Table 1**. In other words for each cell seven rating values were assigned.

The final landslide hazard potential value was obtained by adding the seven values of each cell according to **Table 1**. The values of the failure potential of each cell, which range from 0.5 to 4.4, are classified into three zones of instability. In order to increase slope stability, terrain was divided into high, medium, and low hazard segments. Based on this classification the hazard map was drawn.

4.1. Base Map

There are two maps (**Figure 2** and **Figure 3**) are presented in this section. **Figure 2** is the location map of the

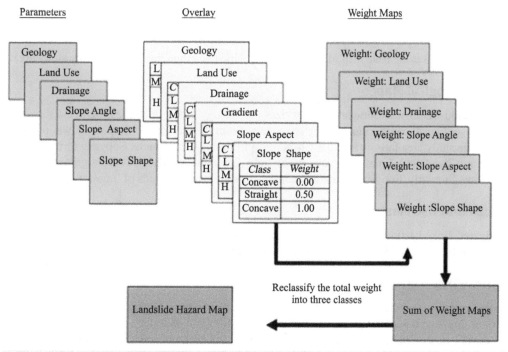

Figure 1. Schematic illustration of superimposition of different layers.

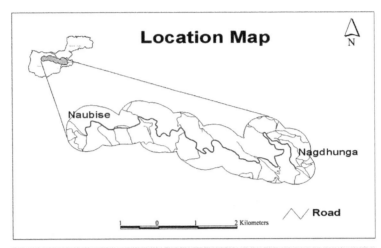

Figure 2. Location map of the study area.

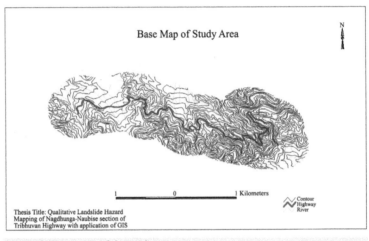

Figure 3. Base map of the study area.

study area. **Figure 3** is the base map of the study area, which is a topographic map of the study area.

4.2. The Geological Map

The Kathmandu Geological Complex characterizes the geology of the study area. **Figure 4** shows the proportion of different geological compositions of the study area. From the Geological Map as shown in **Table 2**, the large portion of the study area falls 60% under Chandragiri Limestone, 20% under Sopyang formation and 12% under Colluviums. Chandragiri Limestone consists of light limestone, partly siliceous-argillaceous, fine-crystalline, echinoderms. Similarly, Sopyang Formation consists of slate with thin beds of limestone.

4.3. The Land Use Map

Non-irrigated sloped terraces and forest-covered slopes are the most found landuse type in the study area as shown in **Table 3**. **Figure 5** shows the Land Use map of the study area.

4.4. River/Drainage

Drainage density is an indicator of run-off condition as well as degree of dissection of landscape [3]. The density network of drainage line indicates high runoff and low infiltration and vice versa. High infiltration due to the intense network of discontinuities, high porosity of rocks and soil and intensity of rainfall, increase pore water pressure and decrease the shear strength of the slope [4]. So, ratings for River/Drainage has been developed by

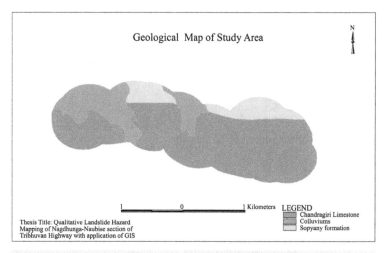

Figure 4. Proportion of different geological compositions of the study area.

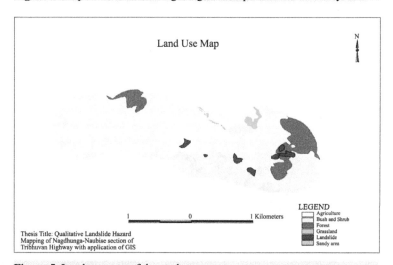

Figure 5. Land use map of the study area.

Table 2. Details of the geology of the study area.

SN	Geology	Rating	Area (Km^2)	Remarks
1	Chandragiri Limestone	0.6	3.69	Highest Composition
2	Sopyang Formation	0.5	0.67	
3	Colluviums	0.00	1.09	

Table 3. Details of land use of the study area.

SN	Land Use	Rating	Area (Km^2)	Remarks
1	Agriculture	0.6	4.41	Highest Area
2	Forest	0.5	0.51	
3	Grassland	1.0	0.05	
4	Sandy Area	0.5	0.01	Lowest
5	Sush and Shrub	0.6	0.40	
6	Landslide	1.0	0.07	

Mountain Risk Engineering (MRE) II handbook and engineering judgment. **Figure 6** shows the river/drainage system of the study area.

4.5. The Slope Class Map

The enclosed slope class map depicts the assigned slope classes and their distribution in the study area. The largest portion of very gentle slopes between 0 to 20 degree is around 36%, followed by areas with inclination of 40 - 90 degree, which makes up a portion of 26%. **Figure 7** shows the different proportions of the slope angles and **Figure 8** shows the different proportions of slope shapes, which are shown below.

4.6. The Slope Aspect Map

The slope exposure was determined by dividing the area into 9 aspect-classes. Each class encloses an angle of 45 degrees. As presented in **Table 4**, the largest part of the area belongs to the flat aspect class (2.18 Km^2) and the smallest to the East aspect (0.06 Km^2). This may be due to the fact that the study area is confined to 500 m on either side of the road section. **Figure 9** shows the proportions of different slope aspect of the study area.

4.7. Final Hazard Map

The percentage distribution of hazard zones and the relative landslide frequency indicates that the low class with 32%, medium class is most found with 51%, and high class with 17% as shown in **Table 4**—Landslide distribution

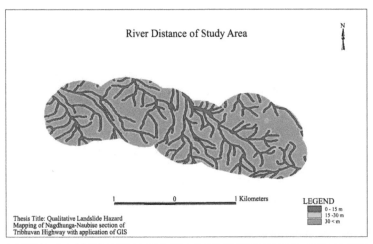

Figure 6. River/drainage system of the study area.

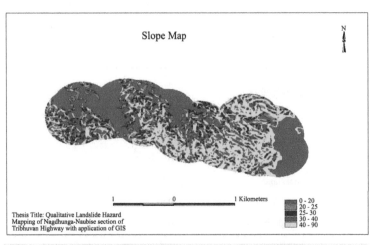

Figure 7. Proportion of different slope classes of the study area.

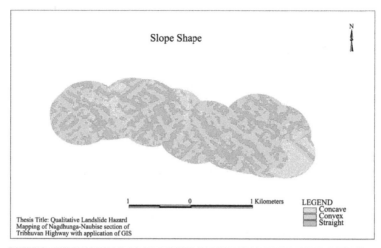

Figure 8. Proportion of different slope shapes of study area.

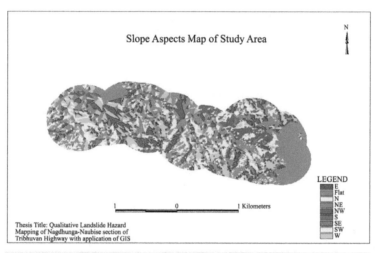

Figure 9. Proportion of different slope aspects of the study area.

Table 4. Details of slope aspect of the study area.

SN	Aspect	Rating	Area (Km²)	Remarks
1	Flat	0.00	2.18	Highest Area
2	N	1.00	0.59	
3	NE	1.00	0.24	
4	E	0.50	0.06	Lowest
5	SE	0.00	0.09	
6	S	0.00	0.33	
7	SW	0.50	0.52	
8	W	0.50	0.70	
9	NW	1.00	0.75	

and percentage distribution of hazard zones and **Figure 10**—Final Landslide Hazard Map of the study area. Another important finding is that most of the present landslides fall under the high hazard zone of the final landslide hazard map (**Figure 10**). This could indication the reliability of the hazard map.

Table 5 also shows the number of landslide counted per km^2, which is a hazard potential unit. To obtain comparable values, landslide percentages were calculated per km^2 hazard segments.

It should be noted that due to the small study area (about 6 km^2) as well as small number but large size of landslides, the calculated landslide frequency is not truly representative.

5. Conclusions

The final hazard map shows that in the areas with hazard potential class "low" with 32% portion, "Medium" is most found with a portion of 51% followed and "high" with the portion of 17%.

The statistical analysis from the calculation of the landslide percentage per Km^2 in the area is 1.1 per Km^2 (highest), which is nearly equal to the value of Middle Mountain region as 1 landslide per Km^2, as suggested by Wagner, 1983.

Due to the lack of important terrain information such as weathering effect, structural discontinuities, seismicity and hydrogeology were not included in the analysis. Rainfall is the main cause of failure, but it is not incorporated in the hazard rating due to the small area and the fact that it is very difficult to keep record of such variation within the small area.

The study characterizes the steep slopes: steep flow channels and fragile geology. These factors have contributed to frequent slope movement and intense erosion process during the heavy rainfall during monsoons. The developed linear infrastructure has also increased the instability of the corridor. Potentially unstable slopes were found to be high in number among steep slopes. Interestingly, the slope movement phenomenon is high in the forest as opposed to the cultivated land. The natural factors seem to have played predominant role in causing slope failures.

To obtain the slope failure potential map, data layers for each factor have been produced and superimposed. It should note that some of these superimpositions could cause the exception. In other words, it is more complex and changeable when many factors appear concurrently.

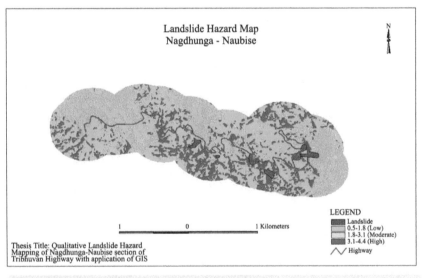

Figure 10. Final Landslide Hazard Map of the study area.

Table 5. Landslide distribution and percentage distribution of hazard zones.

Failure Potential	Area (Km^2)	Area (%)	No. of Landslides	Landslide per Km^2	Landslide Percentage per Km^2
Low	1.70	32	3	0.55	5.87
Medium	2.77	51	2	0.37	9.36
High	0.98	17	6	1.1	3.11
Total	5.45	100	11	2.02	18.34

6. Recommendation

The study of hazard mapping is very useful for the mountainous country like Nepal. This will greatly help to design and construct infrastructures such as road, water supply, and hydropower projects. The application of GIS in preparing hazard mapping is even unique. This gives the ability to capture, manage, manipulate, analyze, model, and display spatially referenced data. So, it is recommended to further expand this type of study by using the technique to a larger area, which can also include critical contributors of landslide such as rainfall, weathering condition, and hydrology. Simple qualitative mapping (pre-assigned rating) method was used to produce the hazard map of the study area. In order to see the level of the accuracy of this method, it is recommended to use other methods statistical (bivariate and multivariate) and deterministic method to compare result.

References

[1] Tianchi, L., Chalise, S.R. and Upreti, B.N. (2001) Landslide Hazard Mitigation in the Hindu Kush-Himalayas. International Center for Integrated Mountain Development, Kathmandu.

[2] Wagner, A. (2000) Slope Stability Mapping of the Chandisthan Sub-Watershed, a Catchment Area of the Marsyangdi River. Halvetas/Swiss Technical Corporation Nepal, Kathmandu.

[3] Ghimire, M. (2002) Geo-Hydrological Processes and Their Impact on the Environment and Socio-Economy of a Watershed Mountain Development, Kathmandu, Nepal. International Center for Integrated Mountain Development, Kathmandu.

[4] Deoja, B., Dhatal, M., Thapa, B. and Wagner, A. (1991) Mountain Risk Engineering Handbook Part I—Subject Background. International Center for Integrated Mountain Development, Kathmandu.

[5] Deoja, B., Dhatal, M., Thapa, B. and Wagner, A. (1991) Mountain Risk Engineering Handbook Part II—Application. International Center for Integrated Mountain Development, Kathmandu.

[6] Deoja, B., Dhatal, M., Thapa, B. and Wagner, A. (1991) Risk Engineering in the Hind Kush—Himalaya. International Center for Integrated Mountain Development, Kathmandu.

[7] Department of Water Induced Disaster Prevention, Government of Nepal (2004) Report of Photo Monitoring in Kathmandu-Naubise Road. Kathmandu.

[8] Ministry of Home (2003) Annual Report. Kathmandu.

[9] Moser, M. (2002) Geotechnical aspects of landslides in the Alps. In: Rybár, J., Stemberk, J. and Wagner, P., Eds., Landslides, 24-26 June 2002, A.A. Balkema, Prague, 23-44.

[10] Sarkar, S., Kanungo, D.P. and Mehrotra, G.S. (1995) Landslide Hazard Zonation: A Case Study in Garhwal Himalaya, India. Mountain Research and Development, 15, 301-309.

[11] Schuster, R.L. and Fleming, R.W. (1986) Economic Losses and Fatalities due to Landslides. Bulletin of Association of Geologists, 23, 11-28.

[12] Sharma, C.K. (1988) Natural Hazards and Man Made Impacts in Nepal Himalaya. Puspha Sharma Publication, Kathmandu.

4

Mapping of Flood Prone Areas in Surulere, Lagos, Nigeria: A GIS Approach

Chidinma Blessing Okoye[1], Vincent Nduka Ojeh[2]

[1]Department of Geography, University of Lagos, Akoka, Nigeria
[2]Wascal, Department of Meteorology, Federal University of Technology, Akure, Nigeria
Email: drojehvn@hotmail.com

Abstract

Increase in the frequency of flooding incidents all over the world and indeed in Nigeria is a major cause for concern. The purpose of this research work is to identify the factors that lead to flooding; and using Geographic Information Systems, map the flood prone areas in Surulere. The causal factors of flooding in Surulere, such as high rainfall intensity and duration, land use patterns, human actions, urbanization, soil impermeability to mention but a few were identified. The pattern of rainfall in Surulere was analysed using 20 years rainfall data from The Nigerian Meteorological Agency. The land use/land cover map as well as the DEM of Surulere was generated in the ArcGis environment with ikonos imagery and 1:50,000 toposheet of Lagos S.E which covers Surulere. The Flood prone areas in Surulere were then mapped. Results show that majority of the land area of Surulere is prone to flooding. Flood mitigation such as early warning systems and flood forecasting, proper urban development, awareness and research should be implemented by the Lagos State Government.

Keywords

GIS, Flooding, Surulere, DEM

1. Introduction

Ojeh and Ugboma [1] asserted that flooding is arguably the weather-related hazard that is most widespread around the globe. Flooding is an event which occurs when there's an overflow of water that submerges land. Flood is one of the natural environmental hazards ravaging the landscape of mankind over the years and whenever flood occur, they result in the loss of properties, lives, destruction of farmlands etc. in most towns in the world [2]. According to [3] flood is an extreme weather event naturally caused by rising global temperature which results in heavy downpour, thermal expansion of the ocean and glacier melt, which in turn results in rise in sea level, thereby causing salt water to inundate coastal lands. Floods or flood waters are temporary inundation of normally dry land areas from the overflow of inland or tidal waters, or from the unusual and rapid accumulation or runoff of surface waters from any source onto lands that are used or usable by man and not normally covered by water [4].

Flooding incidences are becoming a more frequent occurrence in Nigeria. Between 2011 and 2012, there were a number of reported cases of flooding in several parts of the country. The major floods that overtook most parts of Kogi, Delta and Bayelsa states and Onitsha in 2012 is an example (See **Figure 1** below). Areas around the River Niger were totally submerged by floods and over 600,000 residents were rendered homeless, farmlands lost and many killed [5]. Like [6] rightly said "In the year 2012, Nigeria witnessed the highest flood disaster in 100 years, where over ten states of the Federation came greatly under water". According to experts, the floods were caused by excess rainfall which resulted in the over flooding of Rivers Niger and Benue and their tributaries, from Taraba to Adamawa all the way to the southern states of Nigeria [5]. This incident was predicted by The Nigerian Meteorological Agency.

In Lagos, it has become a normal phenomenon for floods to accompany heavy and or prolonged rains. The July 10th flood of year 2011 is perhaps an incident that Lagosians would not forget in a hurry, as there were many devastating effects of the flood. This year also, there have been cases of severe flooding in Lagos after a heavy downpour of rain. On June 28th 2012, Lagos residents were enveloped by floods resulting from a heavy downpour of rain which started the night before and lasted for several hours. A number of houses and roads were submerged by the floods and some people lost their lives. Sections of Lagos-Abeokuta expressway were cut off on both sides by the flood and a portion of the Murtala Mohammed International airport road by Mobil Filing station was submerged. The worst hit areas were Okokomaiko, GRA Ikeja, Ipaja-Ayobo, Shogunle, Apapa, Shomolu, Magodo, Ejigbo and Surulere areas of Lagos state (see **Figure 1** and **Figure 2**). **Figure 2** shows a flooded street in Aguda, Surulere after a rainfall event.

Flooding is a big problem in Surulere, even during the mild rainfalls streets are flooded and many times water rises to house levels. Intensity of rainfall in short period, in rainy season, leads to extremely high runoffs and floods. Surulere is a lowland which makes the situation even worse because the lowland nature hinders water discharge to the sea. Also, due to poor soil infiltration only a small proportion of rainwater seeps into the ground [7]. Because of poor infrastructure planning, buildings often block natural watercourses and canals are too narrow to convey rainwater away from the area. Roads are often unpaved and the hard rain makes them muddy and bumpy. Inhabitants of the LGA wish to have expanded canals, paved roads and better drainage to prevent flooding of homes and other problems during the rains.

The incessant flooding incidents in Lagos and hence Surulere have become a major cause of concern especially because of its disastrous aftermath. At the onset of the rainy season in 2013, The Lagos State government indicated that Surulere LGA is one of the parts of Lagos that are in danger of being flooded. This frequent incident has been linked to changes in the precipitation pattern in area with more rains with higher frequency and

Figure 1. Aerial photograph of flood submerged area in Kogi [8].

Figure 2. A flooded Ijewere street in Aguda, Surulere [9].

intensity. Also, the relief of Surulere, (which is relatively lowland) contributes as it prevents water discharge to the sea. Due to the poor infrastructural planning, buildings often block natural water drainage systems and roads constructed without drainage prevents rainwater from flowing out. Thus, it has become necessary to identify the immediate and indirect causes of flooding in Surulere, and map out the parts of the LGA that are more susceptible to flooding using GIS, so as to proffer solutions to the flooding situation in Surulere.

2. Conceptual Issues and Related Literature

This study hinges on the GIS concept. Geographic refers to anything that relates to space or the arrangement (location and distribution) of objects in space. It implies an interest in the locational identity of any object on, under or above the earth's surface. Information are facts resulting from the processing or refining of raw data. Such facts are usually meaningful, hence of value to some users and could be used for decision making. Geographic information is any information or data that has geographic identity, *i.e.* its specific location or in relation to the surface of the earth is known. A system is an organised assemblage or collection of interrelated components (subsystems), which harmoniously interact with one another towards accomplishing some desired results or pre-defined goals.

A GIS is thus an orderly assemblage of computer-based hardware, software, geographically referenced data, procedures, and human ware (personnel) configured to handle all forms of spatial data to satisfy the geographic information needs of a user [10] According to [11], the field of geographic information systems (GIS) is concerned with the description, explanation, and prediction of patterns and processes at geographic scales. GIS is a science, a technology, a discipline, and an applied problem solving methodology. Simply put, a GIS is an application used to deal with or process spatial information on the computer [12].

Like every other system, A GIS consists of some distinct but fundamentally related components parts. They include hardware, software, data, procedure and personnel or human ware. The hardware refers to the physical computer system and associated accessories necessary for running a GIS software as well as the capture, storage, manipulation and output of spatial data. A typical GIS software often contains tools in form of algorithms (programs, rules or commands) for performing certain tasks as data input, storage, retrieval, analysis, query, output and updating. GIS data refers to facts or information about some geographic feature and includes both geographic and attribute data. There are several procedures involved in a GIS and they include data acquisition, input, storage, sorting, indexing, retrieval, analysis, output and updating, along with the process of decision making. GIS personnel or human ware refer to a group of experts who one way or use or promote the use of GIS and allied facilities such as manufacturers of GIS hardware and software, GIS data providers, managers, users of GIS technology, consultants and so on.

There is a large range of application areas of GIS. They include topographic base mapping, socio-economic and environmental modelling, global (and interplanetary) modelling, and education. Applications generally set out to fulfill the five Ms of GIS *i.e.* mapping, measurement, monitoring, modelling and management [11].

Reference [13] stated that "Apart from traffic congestion, flood is the most common serious physical urban problem in most Nigerian cities". According to them, flooding usually results from high river levels, concentration of overland flow following heavy rainfall, limited capacity of drainage systems and blockage of waterways and drainage channels. Reference [1] observed that flooding in Abraka, the host town of Delta State University, has resulted to the inaccessibility of some roads, destruction of properties, lateness to school, forced migration/relocation and traffic congestion in the area and therefore has negative effect on socioeconomic activities leading to decrease in the productivity of the people.

In their study on natural hazard and crop yield, [2] stated that there was a high level of inundation in farmlands in Oleh soon after every rain event. A mean value of 0.608 kilometres of flooded portion was recorded at Odah in 2011, 0.441 kilometers at Iwhreotah in 2011, 0.547 kilometres at Erorin in 2011 and 0.485 kilometres at Odah in 2012, 0.425 kilometres at Iwhreotah in 2012 and 0.598 kilometres of flooded portion in 2012 at Erorin quarters were generally observed in the farm lands. The study revealed that all the crops cultivated in the area (cassava, melon, yam, maize, plantain) were affected by flooding above 50 percent of total yield of each crop cultivated in the area except yam (46.9%).

[14] wrote on the Environmental Impact of flooding on Kosofe LGA of Lagos State. They stated that the causes of flood tend to vary from one locality to another depending on the available protection and management process. Urbanization and or the concentration of settlements have continued to raise the flood damage, as settlements continue to encroach on the flood prone areas. They also indicated that over reliance on safety provided by flood control infrastructure such as levies, dykes, reservoirs, dam can also result in flood disaster. For example dyke though a flood protective structure can collapse, thereby resulting into immense water destruction. Human action also causes flooding; there is the tendency to encroach on flood plains which attract development due to their flatness, soil fertility and proximity to water. According to them, other causes of increasing flood risk include increase in the proportion of impervious area; deforestation and channel interference e.g. channel suffocation by solid waste.

[15] discussed the negative social impacts of flooding. According to the article, flooding of areas used for socio-economic activities produces a variety of negative impacts. The magnitude of adverse impacts depends on the vulnerability of the activities and population and the frequency, intensity and extent of flooding. Some of the negative impacts highlighted are: Loss of lives and property, loss of livelihood, decreasing purchasing and production power, mass migration, psychosocial effects, hindering of economic growth and development and political implications.

[16] stated as thus—"High rainfall and consequent floods are recurrent phenomenon in India causing loss of lives and damage to livelihood, property, infrastructure and public utilities. However, these events also provide opportunities of positive effects". The positive impacts of floods highlighted include ground water recharge, enhanced water availability in dams/reservoirs, and deposition of fertile silt on flood plains. Surulere is yet to be given adequate research attention for future planning purposes, thus, this study is geared towards making contribution to existing body of literatures from the GIS mapping perspective to flooding in Surulere.

3. Methodology

3.1. The Study Area

Surulere is a local government Area of Lagos State. It is located on the Lagos mainland between 6°05'N and 6°30'N of the equator and 3°21'E and 3°35'E of the Greenwich meridian (**Figure 3**). Surulere is a residential and commercial area, with a land area of 27.05 sq. km. It is part of Metropolitan Lagos. The boundary of Surulere LGA is as follows: On the North, the road that separates Ransome Kuti School through Bishop Street to include Obele-Oniwala up to boundary with Itire at Mushin Local Government; South—the roundabout to Costain to include the whole Abebe village and Eric Moore; East—the railway crossing at Yaba to exclude Railway compound through the swamp at Alaska estate and West—Nepa's high-tension cable on the other side of Jubril Martins through Masha to include both sides of Adelabu street, old census office to Babs Animashaun. Therefore, it can be said that Surulere is bounded on the north by Mushin and Shomolu LGAs, on the South by Ajeromi-Ifelodun and Apapa, on the West by Oshodi-Isolo and Amuwo-Odofin and on the East by Lagos Mainland.

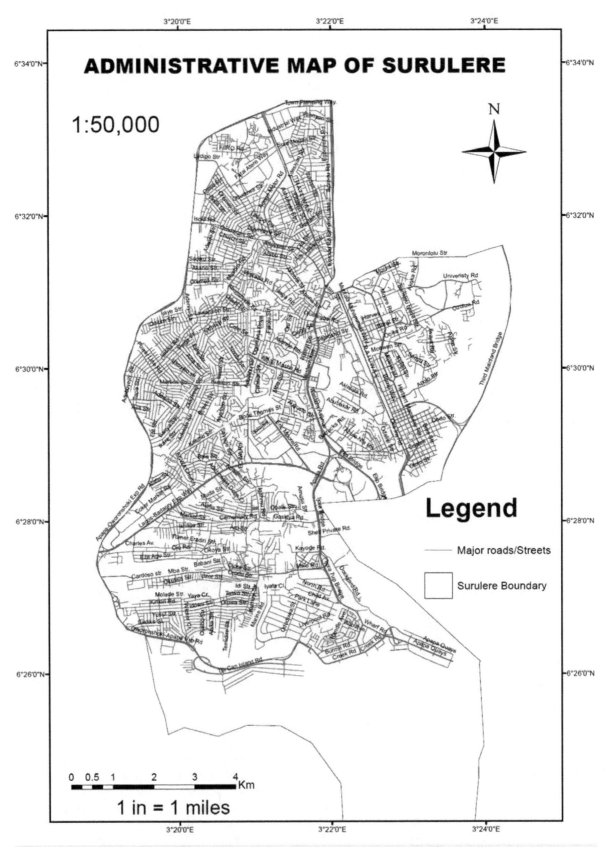

Figure 3. Administrative map of Surulere.

However, for the purpose of this work, the immediate environs (surroundings) of Surulere was considered *i.e.* Oshodi, Mushin, and Lagos Island axis for a holistic view.

Surulere has a tropical wet and dry climate (Köppen climate classification Aw) that borders on a tropical monsoon climate (Köppen climate classification Am). It has a tropical savanna climate that is similar to that of the rest of southern Nigeria. There are two climatic seasons in a year—The rainy season, which lasts from April to October and the rainy season, which lasts from November to March. The heaviest rains fall from April to July and a weaker rainy season occurs in October and November. The peak months of the rainy season are the months of June and July. There is a brief relatively dry spell in August and September and a longer dry season from December to March. The mean monthly rainfall between May and July is about 300 mm while in August and September it is down to 75 mm and in January, it is as low as 35 mm. Generally, however, the mean monthly rainfall through the year is about 120 mm, while the total annual rainfall is almost 1500 mm.

The main dry season is accompanied by winds from the Sahara Desert which between December and early February can be quite strong. The average temperature in January is 27°C and for July it is 25°C. On average the hottest month is March with a mean temperature of 29°C while July is the coolest month but still there doesn't exist much of the hot months. The dominant vegetation in Surulere is tropical swamp forest, comprising fresh waters and mangrove swamp forests.

Surulere is a generally low land covered by nearly uniform terrain made up of mainly sandy soil. The geology consists of quaternary alluvial deposits such as ferralitic red-yellow soil, red-brown, grey and sandy clays, silt, sand gravels and cowrie shells amongst other detrital materials. In slightly elevated upland areas where dry patches occur, the rock type is sedimentary basement complex of Precambrian origin with deposits of rare tainted lignites. The average elevation of Surulere is about 10 m above sea level. Although there are no major water bodies in Surulere LGA, the major water bodies in Lagos state are the Lagos and Lekki Lagoons, Yewa and Ogun rivers. Rivers flowing to the sea form swampy lagoons like Lagos Lagoon behind long coastal sand spits or sand bars. Some rivers, like Badagry Creek, flow parallel to the coast for some distance before exiting through the sand bars to the sea.

Surulere LGA, as of the 2006 census has an estimated population of 1,274,362 people, with 698,403 males and 575,959 females. With a land mass of 27.05 sq. km, the population density is 47111.35 people per sq. km. The population of Surulere is projected to increase to 1,639,572 by 2014 with a population density of 60,613 and 1,692,038 by 2015 with a population density of 62,552; using an annual growth rate of 3.2% [17].

Surulere is a residential and commercial area, with majorly residential areas and a few commercial centres. The Ojuelegba area of Surulere is a major hub of commercial activities, with the former Tejuosho market situated along Ojuelegba road towards the railway axis. The Aguda and Iponri markets are also major markets in the LGA. Streets such as Adeniran Ogunsanya Street, Akerele Street, Ogunlana Drive are major streets where commercial activities also take place. The National and Lagos State Stadiums are both situated in Surulere and serve as sport recreational centres.

3.2. Research Instrument/Data

Primary data

1) 1:50,000 topographic map of Lagos S.E covering Surulere Local Government.

2) A map of Surulere, showing the boundary.

3) Ikonos satellite imagery with resolution of 1 m covering Surulere LGA (see **Table 1** for Specifications of the Ikonos Imagery).

4) 20 years (1991-1998; 2000-2011) monthly rainfall data of Oshodi met station from NIMET. Data for 1999 was not available.

Secondary Data

1) Available data on past recorded flooding incidents in Lagos.

2) Field surveys.

Hardware and Software

1) Hp 630 notebook personal computer with 2 gigabyte RAM, 64-bit operating system and 283 gigabyte local disk memory

2) HP Office jet 5600 series printer

Table 1. Specifications of the Ikonos Imagery.

Spectral Resolution	Im
Spectral Range	Band 1 (blue) 0.445 - 0.516 micrometers Band 2 (green) 0.506 - 0.595 micrometers Band 3 (red) 0.632 - 0.698 micrometers Band 4 (near infrared) 0.757 - 0.853 micrometers
Radiometric Resolution	11 bits per pixel
Temporal Resolution	3 days
Datum	WGS84
Sun Angle	>15°

3) Arc.GIS 10.0 software
4) Envi 4.7 software

3.3. Identifying the Factors That Contribute to Flooding in Surulere

In identifying the factors that contribute to flooding in Surulere, field surveys were carried out by going to a few selected parts of Surulere (Ground truthing).

The rainfall data collected for (Ikeja met station) was analysed to see the rainfall trend in the study area over the years. This was backed up by literature research and the results of the analysis carried out.

3.4. Extraction of the Surulere Map from the Lagos S.E Toposheet

- The ArcGis 10.0 software was launched.
- The 1:50,000 toposheet of Lagos S.E, sheet 279 S.E. of the Nigerian topographic sheet series (Federal Surveys Nigeria, 1966), which was in raster format was then brought into the GIS environment by clicking the add data icon and importing the map from the folder it was saved. The toposheet is shown in **Figure 4**. The boundary map of Surulere was also imported into the GIS environment.
- The maps and the data frame were spatially referenced using the World Geographic Coordinate System (WGS) 1984 referencing system.
- The Lagos S.E toposheet and the Surulere Boundary map were then georeferenced by launching the geo-referencing toolbox and using the add control points tool to pick out points on the 4 edges of the map and inputing the coordinates, which is then saved by clicking rectify on georeferencing dialogue box and clicking update georeferencing.
- The arc catalogue was launched and the editor toolbox was launched.
- A file geodatabase was created and the imported Surulere boundary map was digitized, by using the edit tool in the editor toolbox to pick out points round the boundary, which was then saved.
- The digitized boundary map was then overlayed on the Lagos S.E toposheet.
- The ArcToolbox was launched and the Surulere toposheet was extracted by clicking on spatial analyst tools under the ArcToolbox → extraction → extract by mask → input raster (the Lagos S.E toposheet) → input raster or feature mask data (boundary) → output raster → ok.

3.5. Generation of a Relief Map of Surulere

- Two (2) File Geodatabase layers were created from the ArcCatalog, one for point data (Spot height) and the other for linear data (Contour line).
- The boundary map is overlayed on the extracted toposheet.
- The editing tool was used to digitize the spot height and the contour lines and the values recorded in the attribute table accordingly.
- The symbols are then adjusted accurately.

3.6. Generation of a Land Use/Land Cover Map of Surulere

- The 1 m resolution Ikonos imagery (with bands 1, 2, 3 and 4 already layer stacked), which was already

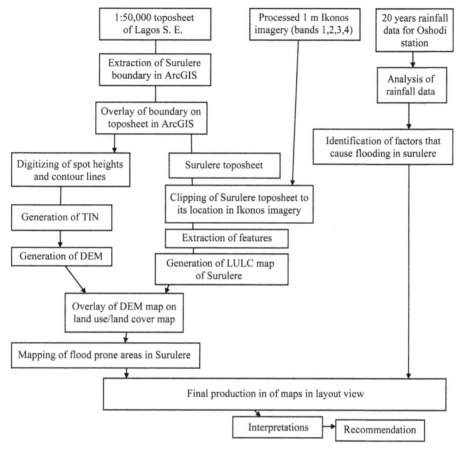

Figure 4. Methodological flow chart of the research design.

processed was imported into the GIS environment.

- The extracted Surulere topomap was clipped to its location on the satellite imagery by going to the Arc-Toolbox → data management tools → Raster processing → clip.
- Layers of Prominent features were then extracted the clipped imagery by vectorization. The features extracted include water bodies, built-up areas, different classes of vegetation, Scattered/open areas.

3.7. Generation of a Digital Elevation Model of Surulere

In creating a DEM of the area, using the contour lines generated, the following steps were taken. (Note the DEM was created using the Triangular Irregular Network (TIN)).

- First, the arc toolbox was launched.
- The 3D analyst extension was activated by clicking on Customize → Extensions → 3D analyst tools.
- The TIN tool was then opened by clicking on ArcToolbox → 3D Analyst tools → TIN Management → Create TIN.
- After clicking on the Create TIN icon, the contour layer was then inputed and the DEM generated: Create TIN → Output TIN → Select output folder → Input spatial reference → Input feature class (the contour layer) → add the height field → ok. (The DEM is generated).
- The colour ramp of the generated DEM is then changed by right clicking on the DEM layer → click on properties → elevation → change the colour ramp.
- The classification method was then edited by clicking on classification (still under properties) → classification method → equal interval → No of classes (the number of classes was changed to 9) → Apply → Ok.
- The DEM was then converted to raster using conversion tools by going to 3D analyst tools → conversion → From Tin → Tin to raster → Input Tin.

3.8. Mapping of Flood Prone Areas in Surulere

- The generated DEM map was overlayed on the Land use/Land cover map.
- All areas below 60ft were mapped as areas more prone to flooding, while areas btw 60 and 100ft were mapped out as areas more prone to flooding.
- Also, built up areas close to water bodies were also considered as areas prone to flooding.

3.9. Production of Generated Maps

All the generated maps were taken one by one to the layout view, by clicking on the layout view icon
- The graticule was inserted by clicking on view → dataframe properties → grid → graticule.
- The other map features such as the neatline, legend, scale text, scale bar, north arrow and title were all included in the various maps by clicking on insert and selecting the features.
- After the maps features were properly placed and positioned, the maps were exported by clicking on file → export map. All maps were saved in pdf format and printed.

4. Results and Discussion

4.1. Factors That Contribute to Flooding in Surulere

The rainfall data analysis carried out shows that over the years, the rainfall amount in the study area has significantly high. There has been some consistency in the rainfall pattern with slight variations. The mean monthly rainfall is 120.2 cm while the mean annual rainfall is 1441.22 cm. (See **Table 2**).

Table 2 shows that the rainfall amounts over the period covered is relatively high. In the **Figure 5** below, it can be seen that over the years, Surulere has experienced a steady rise in annual rainfall, although there was a decrease between year 1995 and 2000. However, there has been a steady increase ever since and the trend is most likely to continue due to global warming.

The Field Survey carried out shows that a lot of houses in Surulere are constructed on drainage channels. Also, some roads are constructed without adequate drainage channels for water flow. Some of the drainage channels available are blocked due to continuous dumping of refuse in the channels e.g. in Cele and Itire areas.

Figure 6 displays the topography of the study area. The relief and DEM maps generated show that Surulere is a relatively flat land, with the highest elevated peak at 100ft (See **Figure 7** and **Figure 8**). Majority of the land area is below 50ft (15 m); some areas are even as low as 30ft. Therefore, it can be classified as a lowland, thus making it difficult for accumulated water from rainfall to flow to the lagoons and creeks.

The land use/land cover map generated (**Figure 9** and **Figure 10**) shows that majority of the study area is

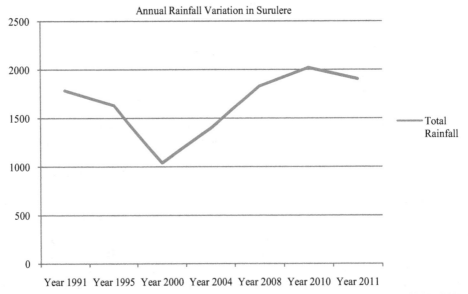

Figure 5. Annual variation in rainfall totals in the study area.

Table 2. 20 years monthly rainfall data (in cm) for Oshodi met station (Source of data: NIMET, Oshodi).

OSHODI	1991	1992	1993	1994	1995	1996	1997	1998	2000	2001
JAN	TR	0	TR	51.6	TR	8.5	TR	TR	TR	7.5
FEB	13.6	TR	53.8	8.3	73.4	108.6	0	33.3	33.3	4.4
MAR	69.7	44.7	111.3	56.3	165.3	94.6	126.5	20.3	20.3	12.9
APR	288.3	84.1	187.5	591	168.7	241.5		54.8	54.8	212.5
MAY	224.7	426.4	252.7	145.1	236	176.7	223.8	114.7	114.7	206.9
JUN	424.8	278.6	273	330.9	287.2	223.1	427.4	204.9	204.9	214.6
JUL	357.1	144.6	118.9	80.8	280.4	301.9	70.5	87.3	87.3	101.5
AUG	41.5	16.4	28.1	28.6	98.9	170.5	120.1	93.6	93.6	24.1
SEPT	215.7	162.4	163.5	139.4	108.5	125.3	123.3	291.8	291.8	199.4
OCT	143.3	100.3	110.8	154.7	151.2	nil		83.8	83.8	108.5
NOV	7.7	69.2	94.9	6.4	21.9		56.8	38	38	12
DEC	0	21.4	9.8	10.2	40.9			17.8	17.8	13.3
ANNUAL	1786.4	1348.1	1404.3	1603.3	1632.4	1450.7	1148.4	1040.3	1040.3	1117.6
MONTHLY	148.87	112.34	117.03	133.61	136.03	120.89	95.70	86.69	86.69	93.13
MAX	357.1	426.4	252.7	330.9	287.2	301.9	427.4		291.8	214.6
MAX MNT	JUL	MAY	MAY	JUN	JUN	JUL	JUN		SEPT	JUN
MIN	0	0	TR	6.4	TR	8.5	0		TR	7.5
MIN MNT	DEC	JAN	JAN	NOV	JAN	JAN	FEB		JAN	JAN

OSHODI	2002	2003	2004	2005	2006	2007	2008	2009	2010	2011
JAN	TR	63.2	10.4	0	81.7	0	1.8	1.3	49.3	0
FEB	34.4	35.3	75	73.1	15.2	1.1	24.2	17.6	48.8	184.4
MAR	28.4	75.2	113.9	56.7	93.6	62.8	85.2	66.8	70.1	5.1
APR	215.1	491.7	143.9	80.6	15.3	19.2	57.5	178.6	139.1	64.4
MAY	129.4	136.2	266.1	226.2	234.8	129.2	178.1	154.1	124.3	226.8
JUN	206.7	241.5	204.8	250.7	177.6	403.3	349.6	462.2	424.5	332.3
JUL	403.3	51.4	47.7	280.7	186.1	235.2	412.8	162.8	112.9	424.2
AUG	43.9	16.6	22.5	3.6	46	212.4	169.7	34.3	276.9	56.9
SEPT	141.7	99.3	286.4	137.7	314.6	107	255.5	88.9	229.5	208
OCT	152.4	77.8	192.8	77.6	83.2	100.9	79.8	200.1	351.9	256.9
NOV	46.5	130.8	30.1	43.5	23.3	27.6	155.3	43.2	138.6	145.2
DEC	25.8	TR	0	21.3	0	23.3	60.6	1.8	55.3	0
ANNUAL	1427.6	1419	1393.6	1251.7	1271.4	1322	1830.1	1411.7	2021.2	1904.2
MONTHLY	118.97	118.25	116.13	104.31	105.95	110.17	152.51	117.64	168.43	158.68
MAX	215.1	491.7	286.4	280.7	314.6	403.3	412.8	462.2	424.5	424.2
MAX MNT	APR	APR	SEPT	JUL	SEPT	JUN	JUL	JUN	JUN	JUL
MIN	TR	TR	0	0	0	0	1.8	1.3	48.8	0
MIN MNT	JAN	DEC	DEC	JAN	DEC	JAN	JAN	JAN	FEB	JAN/DEC

Mean Monthly Rainfall = 120.01

Mean Annual Rainfall = 1441.22 TR = Trace Amount of Rainfall

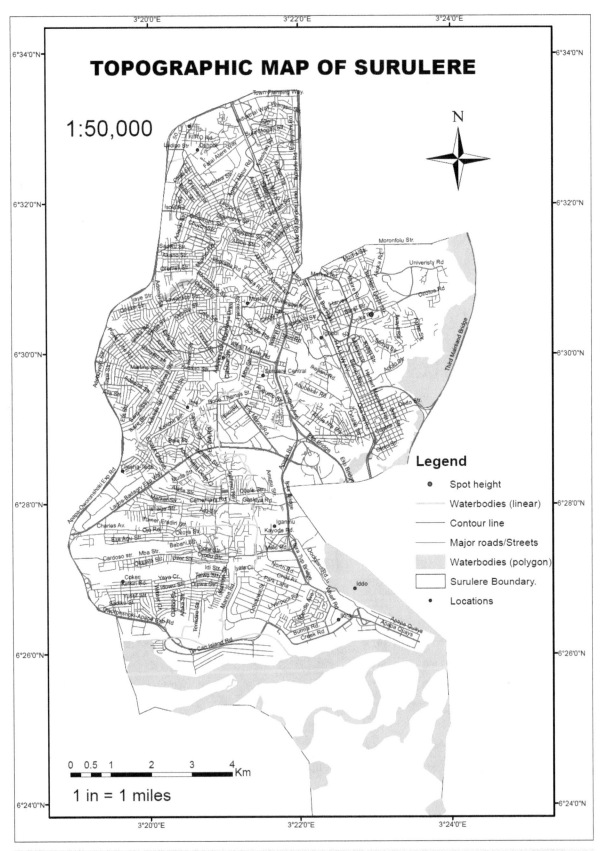

Figure 6. Topographic map of Surulere.

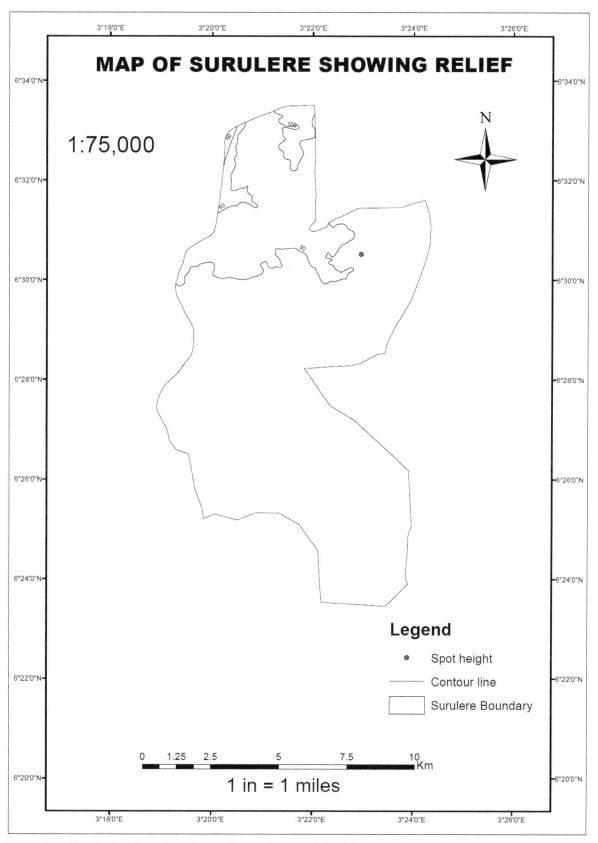

Figure 7. Map of Surulere showing relief.

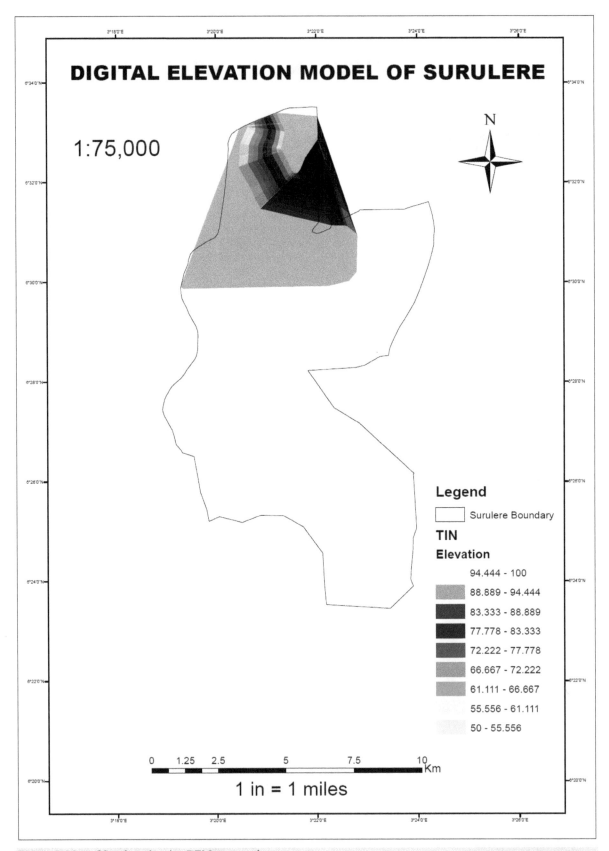

Figure 8. Map of Surulere showing DEM generated.

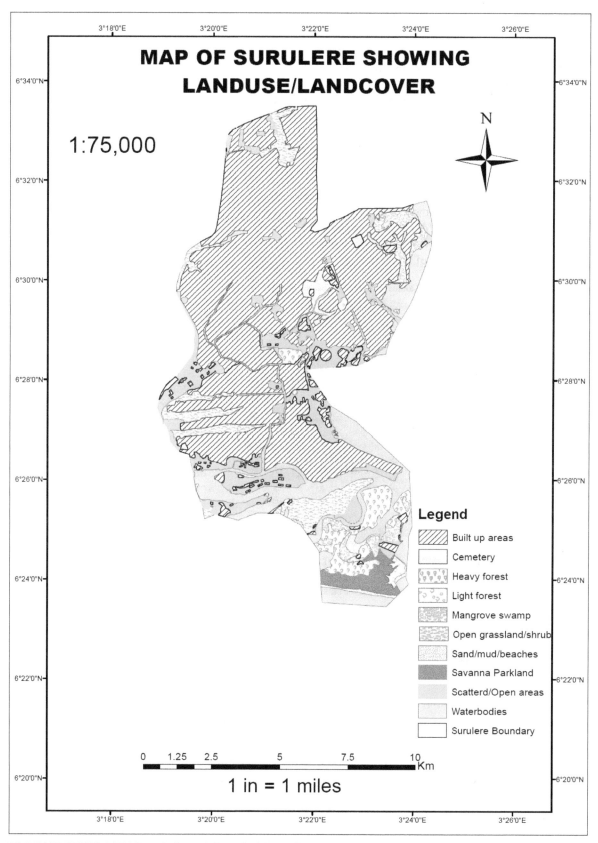

Figure 9. Map of study area showing land use/land cover classification.

built up. The area doesn't have sufficient drainage channels. A lot of the buildings are either constructed along drainage channels or blocking drainage paths.

The pie chart above shows that a major percentage of the study area is built-up.

From the **Table 3**, it can be seen that 46.09% of the land area is built up while 24.54% are open/scattered areas. Therefore, most of the land surface is impermeable and rainwater falling cannot be absorbed into the underground soil. There is also a presence of swampy vegetation and water bodies. Most of the features in the built up areas *i.e.* roads, buildings, etc. are constructed along drainage channels, preventing free flow of water thereby causing rainwater to accumulate on the land surface, causing flooding.

Therefore, in summary, the factors that contribute to flooding in Surulere are:

- High rainfall amounts, intensity and duration.
- Majority of the area is built up, with little or no permeable layer.
- Construction of buildings along drainage channels.
- Construction of roads without adequate drainage channels.
- Refuse being dumped in drainage channels.
- Construction of buildings close to water bodies.
- The lowland topography of Surulere.

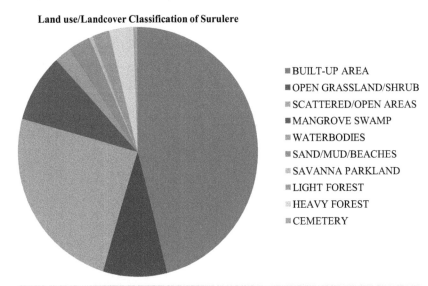

Figure 10. Pie chart showing LULC classification.

Table 3. Land use/land cover classification of Surulere.

LULC Class	Area (sq·m)	Perimeter	% Coverage
Built-up Area	30455.46	37905.67	46.09
Open Grassland/Shrub	5741.4	5398.1	8.69
Scattered/Open Areas	16218.4	15841.08	24.54
Mangrove Swamp	5818.81	2730.4	8.81
Waterbodies	1490.09	2565.61	2.26
Sand/Mud/Beaches	1940.45	4018.32	2.94
Savanna Parkland	322.93	749.56	0.49
Light Forest	1592.27	1750.53	2.41
Heavy Forest	2166.16	2918.94	3.28
Cemetery	332.84	1312.42	0.50
Total	66078.81	75190.63	100

4.2. Identification of Flood Prone Areas in Surulere

From the analysis carried out, the entire Surulere area can be said to be susceptible to flood due to combined factors mentioned above (See **Figure 11**). If adequate measures are not taken, with the high rainfall amounts and intensity experienced, the area is in danger of being overtaken by flood.

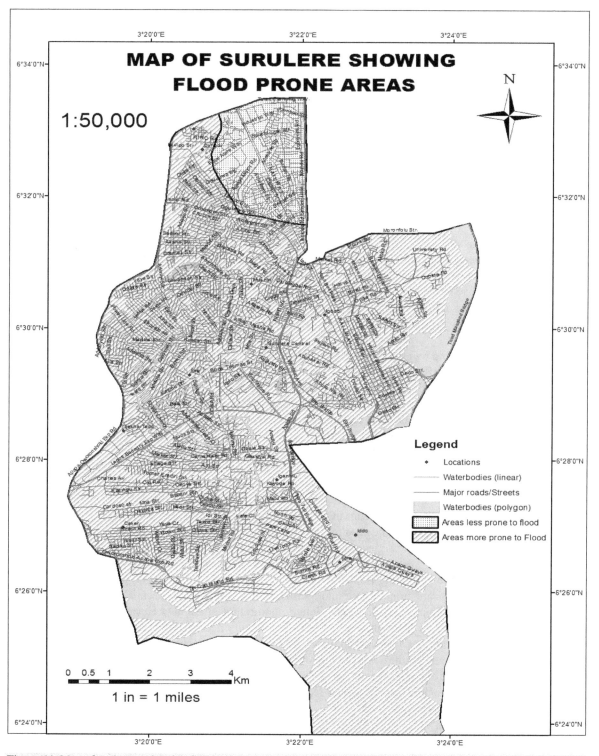

Figure 11. Map of study area showing flood prone areas.

5. Conclusion and Recommendations

The purpose of this research work is to map the areas prone to flooding in Surulere. From the results of the analysis carried out, it is clear that the entire Surulere area is prone to flooding. This is due to a number of factors mentioned in the paper. If serious measures are not taken soon, taking the increasingly high rainfall into consideration, the area is in danger of being submerged with floods with lives and properties in danger of being destroyed. The year, 2014 alone, high rainfall intensity has led to flood incidents that left washed out roads, destroyed structures and loss of lives in its trail e.g. in the Ebute-metta, Cele and Aguda areas. Therefore, it is important for the Lagos State government to put in place measures to mitigate flooding in Surulere. Such measures include:

- Careful planning of development in potentially flood prone areas. This should be supported by appropriate legislation, public information and education programmes (to ensure that residents understand the flood risk), flood insurance, and flood warning systems (to reduce the impact of floods).
- Flood forecasting and flood warning services. This service has been readily provided by the Nigerian Meteorological Agency.
- Roads should be constructed with proper drainage channels and dumping of refuse in drainage channels should be prohibited and penalties attached to non-adherence.
- More drainage channels and reservoirs should be constructed such as canals and gutters. Side drains and gutters could have removable precast concrete or steel cover for ease of maintenance. Water meadow areas can be created with which to divert flood water.
- Drainage channels should be cleaned out regularly to avoid blockage by silts and mud. This can be achieved by enforcing the monthly environmental sanitation by the local government authorities.
- Runoff control - source control measures should be put in place to reduce the amount of runoff in the event of flooding. These measures include construction of permeable pavements, afforestation and artificial recharge. Also stores of runoff can be created such as wetlands, detention basins and reservoirs [18].
- More in-depth research should be carried out on the impact of global climate change and associated impacts on floods.
- Flood risk maps should be produced and a comprehensive floodplain management plan should be developed for areas likely to be affected by floods. Also flood management studies should be carried out to discover more advanced and effective ways to control flooding.
- The government should put a ban on the construction of buildings on flood plains with penalties attached. However, there should be adequate awareness and dissemination of this information in the media and also the availability of flood risk maps to the public.
- Residents of Surulere, as groups and as individuals, and corporate bodies should be implored and encouraged to embark on some palliative measures such as dredging and re-dredging of drains, erosion passages and others; and construction of embankments and channelization of some routes that are prone to flooding.
- Resettlement strategies and emergency preparedness plans should be developed by the involved the government in order to easily evacuate citizens in the occurrence of flood disasters.

    Reference [18] best summarized the need for emergency plans, stating as follows:

    "Flood warnings and timely emergency action are complimentary to all forms of intervention. A combination of clear and accurate warning messages with a high level of community awareness gives the best level of preparedness for self-reliant action during floods. Public education programs are crucial to the success of warnings intended to preclude a hazard from turning into a disaster. Evacuation is an essential constituent of emergency planning and evacuation routes may be upward into a flood refuge at a higher elevation or outward, depending upon the local circumstances. Outward evacuations are generally necessary where the depths of water are significant, where flood velocities are high and where the buildings are vulnerable. Successful evacuations require planning and awareness among the population of what to do in a flood emergency. Active community participation in the planning stage and regular exercises help ensure that evacuations are effective. The provision of basic amenities such as water supply, sanitation and security in areas where refugees gather is particularly important in establishing a viable evacuation system."

- Finally, Flood recovery measures such as counseling, compensation and insurance should be put in place. The pros and the cons of the policy of flood insurance should however be weighed before implemented. Also, the sociological aspects of coping with flood in affected areas should be taken seriously and appropriate counseling given to affected victims.

Currently, Some of these measure are already being put in place by the Lagos State government. An article in the [19] stated, "Given the prediction of Nigerian Meteorological Agency (NIMET) that this year will witness heavy rains, the Lagos Government has commenced construction of drainages and canals in the Surulere Local Government Area to reduce the impacts on its citizens".

Also, The Lagos State Government is clearing slums, especially those on the drainage channels and right of way. Recently, at Adelabu area in Surulere, behind an elitist school, Fountain Heights Secondary School, a slum community of over 2,000 inhabitants including women and children, were given an eviction notice to allow for the construction of channels and drainage to check flooding in Surulere, which has been a yearly occurrence due to the activities of these people [19]. Also, the Lagos state commissioner for environment, Mr Tunji Bello said that the government would no longer tolerate the construction of shanties and on drainage channels.

However, the governments both at the state and local levels have to do more in order to prevent/reduce future occurrences of flooding in Surulere. If all or most of the measures listed above are taken into consideration and embarked on, flood incidences in Surulere would be greatly reduced.

References

[1] Ojeh, V.N. and Ugboma, P. (2012) Flood Hazards in Urban Niger Delta: A Case Study of Abraka Town, Delta State, Nigeria. *International Journal of Environmental Engineering Research*, **1**, 23-29.

[2] Ojeh, V.N. and Victor-Orivoh, A.F. (2014) Natural Hazard and Crop Yield in Oleh, South-South Nigeria: Flooding in Perspective. *Journal of Earth Science & Climatic Change*, **5**, 181. http://dx.doi.org/10.4172/2157-7617.1000181

[3] Etuenovbe, A. (2011) The Devastating Effect of Flooding in Nigeria. http://www.fig.net/pub/fig2011/papers/ts06j/ts06j_etuonovbe_5002.pdf

[4] Water Words Dictionary (2000) A Compilation of Technical Water, Water Quality, Environmental, and Water-Related Terms. Nevada Division of Water Planning Department of Conservation and Natural Resources, 495.

[5] Nkeki, F.N., Henah, P.J. and Ojeh, V.N. (2013) Geospatial Techniques for the Assessment and Analysis of Flood Risk along the Niger-Benue Basin in Nigeria. *Journal of Geographic Information System*, **5**, 123-135 http://dx.doi.org/10.4236/jgis.2013.52013

[6] Okonkwo, I. (2013) Effective Flood Plain Management in Nigeria: Issues, Benefits and Challenges. Transparency for Nigeria, 2013. http://transparencyng.com

[7] Aina, T.A., Etta, F.E. and Obi, C.I. (1994) The Search for Sustainable Urban Development in Metropolitan Lagos, Nigeria: Prospects and Problems. *Third World Planning Review*, **16**, 201-219.

[8] Ogala, E. (2012) Four People Missing, Hundreds of Farmlands Submerged in Kogi Flood. Premium Times, September 28, 2012. http://www.premiumtimesng.com

[9] Vanguard Newspaper (2012) More Photos of Lagos Flood, by Citizen Reporters. Vanguard Newspaper, June 11, 2012. http://www.vanguardngr.com

[10] Uluocha, N.O. (2007) Elements of Geographic Information Systems. Sam Iroanusi Publications, Lagos.

[11] Longley, P.A., Goodchild, M.F., Maguire, D.J. and Rhind, D.W. (2005) Geographical Information Systems: Principles, Techniques, Management and Applications. John Wiley and Sons Ltd., New Jersey.

[12] Sutton, T., Dassau, O. and Sutton, M. (2009) A Gentle Introduction to GIS. Chief Directorate: Spatial Planning & Information, Department of Land Affairs, Western Cape.

[13] Olajuyigbe, A.E., Rotowa, O.O. and Durojaye, E. (2012) An Assessment of Flood Hazard in Nigeria: The Case of Mile 12, Lagos. *Mediterranean Journal of Social Sciences*, Department of Urban and Regional Planning, Federal University of Technology, Akure, **3**, 367-375.

[14] Oyinloye, M., Olamiju, I. and Ogundiran, A. (2013) Environmental Impact of Flooding on Kosofe Local Government Area of Lagos State, Nigeria: A GIS Perspective. *Journal of Environment and Earth Science*, Department of Urban and Regional Planning, School of Environmental Technology, Federal University of Technology, Akure, **3**.

[15] Associated Programme on Flood Management [APFM] (2013) Urban Floods: What Are the Negative Social Impacts of Flooding? http://www.apfm.info/?p=2459

[16] National Rainfed Area Authority (2011) The Impact of High Rainfall and Floods on Ground Water Resources in the Krishna River Basin (during 1999-2009). National Rainfed Area Authority, Planning Commission, New Delhi.

[17] Lagos Bureau of Statistics (2012) Abstract of Local Government Statistics. Lagos Bureau of Statistics, Ministry of Economic Planning and Budget, Ikeja, Lagos.

[18] WMO (2009) Integrated Flood Management Concept Paper. Associated Programme on Flood Management, World Meteorological Organization, WMO-No. 1047, 2009.

[19] The Guardian Newspaper (2013) Lagos Engages Constructors in Surulere Drainage Project. The Guardian Newspaper, Monday, March 3rd, 2013.

5

Remote Scheduling System for Drip Irrigation System Using Geographic Information System

Kadeghe G. Fue, Camilius Sanga

Computer Centre, Sokoine University of Agriculture, Morogoro, Tanzania
Email: kadefue@suanet.ac.tz, sanga@suanet.ac.tz

Abstract

The Internet is widely accessible in Tanzania. Most of the technologies used in different organizations have changed to address their functions using web based information systems. In this paper, attempt is made to design software system using geographical information system (GIS) for the spatial and temporal distribution of irrigation supply for large-scale drip irrigation systems in Tanzania. Map based information system has gained popularity after evolution of simple tools to present spatial information using Internet. Due to water scarcity, it is envisioned that by 2050 the world won't have enough water for communities, industries and agriculture. Web based precision irrigation system refers to deployment of remotely precision irrigation services using the application interface that connects to the Internet. Hence, this study presents the GIS in the context of precision farming to achieve precision irrigation strategy with special reference to precision farming of tea in Tanzania. The GIS-based irrigation scheduling system was designed for the scheduling daily drip irrigation water deliveries and regular monitoring of irrigation delivery performance for maximum yield. The "Scheduling" program computes the right amount of irrigation deliveries based on tea water requirements. The "Monitoring" program gives information on the uniformity of water distribution and the shortfall or excess.

Keywords

Precision Farming, Irrigation, Scheduling, GIS, Software, System, Remote Scheduling

1. Background Information

Tea yield in Tanzania is primarily limited by insufficient quantity and timing of water and nutrients required for optimum growth. In addition, tea quality is significantly affected by traditional and cultural methods used for tea farms in Tanzania [1].

The quality of Tanzanian tea if increased will have a great impact in the world market compared to the current situation [1].

To compare with Florida, drip irrigation has already proved very successful for optimum production of high value horticultural crops such as grapes, citrus and peaches. On the contrary, current methods and instruments of precision agriculture may not be suitable for Tanzania or they may need modification to apply in Africa after extensive research.

[2] states that stakeholders in Tanzania demand for low-cost irrigation systems, reduced electricity bills and pumping costs and also:

- They need for more efficient irrigation systems which demand less water with view to minimizing depletion of water in reservoirs/dams in response to decreasing amounts of rainfall.
- Stakeholders decision (through the Tea Association of Tanzania) embarked on research on crop water management, including drip irrigation and fertigation of tea. This was due to low availability of water due to poor rainfall during 2003/04-2006/07.
- Saving in water for irrigation by up to 50% is a necessary step for all stakeholders in tea farming.
- Saving in labor for irrigation by 85% is a necessary measure for all stakeholders in tea farming.
- Producing the highest yield, up to 6070 kg/ha is a proposed plan for yield for all stakeholders in tea farming.

Therefore, it's evident that the current decrease in production of tea in Tanzania has been attributed due to unfavorable weather conditions especially poor rainfall in most growing areas. Small-scale farmers of tea need an affordable and effective method to be deployed in their farms at reduced cost or at minimum cost. This can only be attained by a specific irrigation scheduling that is set to optimize the available water resources.

The benefits of having such a system are: reducing monitoring costs, improving the speed of decision making by supporting the decision-makers with real time information, ability to be accessed by everyone and everywhere on the Internet, reducing time and minimizing effort to reach data, high speed, security and capability of having high rate of error handling with new Internet technologies, centralized database that provide a single source of common information for standardization and faster retrieval and selective modification of information [3] [4], and finally the ability to produce reports based on user specified parameters.

2. Objectives

In this study, GIS was deployed to improve irrigation strategy with special reference to precision farming of tea in Tanzania. The GIS-based irrigation system algorithm was developed for the scheduling of daily drip irrigation water deliveries and regular monitoring of irrigation delivery performance. The "Scheduling" program was developed to compute the right amount of irrigation deliveries based on crop water requirements. The "Monitoring" program was developed to give information on the uniformity of water distribution and the shortfall or excess [5]. For optimal results, the developed software should be able to be tested and simulated for irrigation scheduling with allowed water stress depending upon farmers selected irrigation requirements, water restrictions and weather conditions. Hence, the following are the specific objectives under this study:

- To create spatial and weather databases for tea farms.
- To design and develop GIS-based software to schedule irrigation.
- To test and simulate the algorithm of irrigation scheduling as per farmers, selected irrigation requirements, water restrictions and weather conditions.

3. Literature Review

Precision farming aims to manage production inputs over many small management zones rather than on large zones. It is difficult to manage inputs at extremely fine scales, especially in the case of the tea irrigation system. However, in real sense we expect site-specific irrigation approach to potentially improve the overall water management in comparison to irrigated farms of hundreds of acres. A critical element of the irrigation scheduling and management is the accurate estimation of irrigation supplies and its proper allocation for the irrigation of structures based on the actual planted areas. All irrigation scheduling procedures consist of monitoring indicators that determine the need for irrigation. The final decision depends on the irrigation criterion, strategy and goal. Irrigation scheduling is the decision of when and how much water to apply to a field.

The amount of water applied is determined by using a criterion to determine irrigation need and a strategy to prescribe how much water to apply in any situation. The right amount of daily irrigation supply and monitoring at the right time within the discrete irrigation unit is essential to improve the irrigation water management of a scheme [5].

Many computerized tools have been used for scheduling irrigation deliveries and improving the irrigation project management. One such tool is a Geographical Information System (GIS). Its use in irrigation management with their large volumes of spatially and temporally distributed data is most beneficial. The GIS capability to integrate spatial data from different sources, with diverse formats, structures, projections or resolution levels, constitute the main characteristics of these systems, thus providing needed aid for those models that incorporate information in which spatial data has a relevant role [5] [6]. This explained about the capability of GIS for decision-making. The possibility of GIS for easily creating and changing scenarios allows the consideration of multiple alternatives of irrigation scheduling, including the adoption of crop specific irrigation management options. Scenarios may include different irrigation scheduling options inside the same project area applied to selected fields, crops, or sub-areas corresponding to irrigation sectors. This allows tailoring irrigation management according to identified specific requirements [7].

The irrigation scheduling alternatives are evaluated from the relative yield loss produced when crop evapotranspiration is below its potential level. Examples of those successful applications are presented by [7] [8] for surface irrigation in the Mediterranean region.

Irrigation scheduling is the farmers decision process relative to "when" to irrigate and "how much" water to apply at each irrigation event. It requires knowledge of crop water requirements and yield responses to water, the constraints specific to the irrigation method and respective on farm delivery systems, the limitations of the water supply system relative to the delivery schedules applied, and the financial and economic implications of the irrigation practice [9]. Irrigation scheduling models are particularly useful to support individual farmers and irrigation advisory services [8].

[8] reported about an irrigation model embedded within the GIS. The data were correlated to digital data sets on soils, agro-climate, land use, and irrigation practice to produce tabular and mapping outputs of irrigation need (depth) and demand (volume) at national, regional, and catchment levels. The GIS approach allows areas of peak demand to be delineated and quantified by sub-basins. The GIS-based modeling approach is also currently being used to administer irrigation needs for irrigated crops [8] [9]. Map and tabular output from the GIS model can provide licensing staff with the information necessary to establish reasonable abstraction amounts to compare against requested volumes on both existing and new license applications for spray irrigation.

The main value of models results from their capabilities to simulate alternative irrigation schedules relative to different levels of allowed crop water stress and to various constraints in water availability [9].

The main limitation of simulation models is that some model computations are performed at the crop field scale for specific soil, crop, and climate conditions, which characterize that crop field and the respective cropping and irrigation practices. When the computation procedure is applied at the region scale it becomes heavy and slows due to the need to consider a large number of combinations of field and crop characteristics to be aggregated at sector or project scales [7].

Proper irrigation scheduling can reduce irrigation demand and increase productivity. A large number of tools are available to support field irrigation scheduling, from in-field and remote sensors to simulation models. Irrigation scheduling models are particularly useful to support individual farmers and irrigation advisory services [10].

[11] outlined applications of GIS-based modeling. For financial analyses, GIS is used to quantify and map the total financial benefits of irrigation ($/ha) and the financial impacts of partial or total bans on abstraction for irrigation. Irrigation water requirements are determined by mapping the spatial distribution of water requirements based on soil and crop distributions. GIS-based modeling approaches are used to establish irrigation scheduling based on water-balance modeling.

GIS databases for irrigation include coverage for crops, irrigation methods, and soils. These data are coupled with agro-climatic data to provide information on growing-season and water-use requirements.

As previously stated that the economic benefits of having such a system are: reducing monitoring costs, improving the speed of decision making by supporting the decision-makers with real time information, ability to access by everyone and everywhere over the Internet, reducing time and minimizing effort to reach data, high speed, security and high rate of error handling with new Internet technologies, having centralized database that provides a single source of common information which provide standardization and faster retrieval and selective modification of information, and finally, the ability to produce reports based on user specified parameters [3].

4. Materials and Methods

4.1. GIS-Based Irrigation System Scheduling Model

GIS tools and integrated models find extensive application for scheduling of water applications for an irrigation system. The GISAREG model is representative of a GIS-based modeling package directed to improving irrigation scheduling. The model uses the ArcView software and Avenue scripting language to integrate the spatial and attribute databases with legacy irrigation-scheduling models. The model can be applied for different water-management scenarios and produces crop irrigation maps and time-dependent irrigation depths at selected aggregation levels, including the farm scale. GISAREG simulates alternative irrigation schedules relative to different levels of allowed crop water stress as well as various constraints in water availability. The irrigation scheduling alternatives are evaluated from the relative yield loss produced when crop evapotranspiration is below its potential level [7].

GISAREG inputs are precipitation, reference evapotranspiration, total and readily available soil water; soil water content at planting; and crop factors relative to crop growth stages, crop coefficients, root depths, and water-yield response factors. Various time-step computations are possible, from daily up to monthly, depending on weather data availability. GISAREG model results include annual crop irrigation requirements (mm), readily available water at the beginning and end of the irrigation period (mm), total available water in deep soil layers at the beginning of the irrigation (mm), percolation due to excess of irrigation (mm), precipitation during the irrigation period (mm), non-used precipitation (mm), cumulated actual and maximum evapotranspiration (mm), and monthly irrigation requirements (mm). Results are displayed on map and tabular formats.

The proposed system in this study used GISAREG model with some moderation in order to achieve the model that is intended for tea farms. This approach was taken since GISAREG can accommodate more crops but the goal was to develop crop specific scheduling system.

4.2. Soil-Water Content Balance Computations for Tea Farm

The standard method adopted for the calculation of crop water requirements is based on the evaporative demand of the crops for the prevailing stage of growth. There is potential for structuring information to improve the irrigation deliveries, and to develop an information system to improve decision-making in the operation and management of the scheme.

Hydrologic models are valuable tools for water resources management. For irrigation scheduling and crop-water requirement estimation, hydrologic simulation models commonly used the water balance approach [12].

On the other hand, using the soil moisture neutron probe (SMNP) to measure soil water content and determine the optimal irrigation rates and schedules is required to ensure a high yield.

Theoretically, a quantitative estimation of major components of field water balance (**Figure 1**) is important for improving water delivery performance. The main components are humidity, rainfall, temperature, sunshine hours, wind speed, transpiration and evaporation. Analysis of water balance can give management decisions on how the scheme ought to be operated to ensure better distribution of irrigation water to the service areas. The components of the water balance model for a tea field for a given period are as follows:

- MDx is moisture content depth in the field on the xth day, cm;
- MD(x − 1) is moisture content depth in the field on the (x − 1)th day, cm;
- IRx is amount of irrigation water supplied on the xth day, cm;
- ERx is rainfall received on the xth day, cm;
- ETx is crop evapotranspiration on xth day, cm;
- DRx is drainage requirement on the xth day, cm;
- x is irrigation period, day.

Hence, the generalized water balance equation for tea field can be expressed as follows:

$$MDx = MD(x-1) + IRx + ERx - ETx - DRx$$

ETx is the rate of evapotranspiration measured from a large area, covered by green grass, 7 to 16 cm tall,

Hydrological Cycle

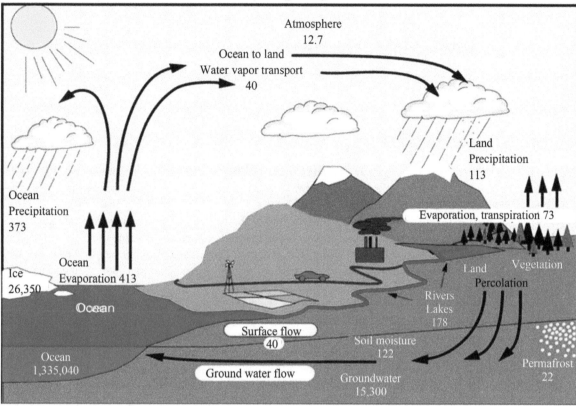

Units: Thousand cubic km for storage, and thousand cubic km/yr for exchanges

Figure 1. Hydrological cycle (https://waterdropblog.files.wordpress.com/2008/06/01_hydrologicalcycle.gif).

which grows actively, completely shades the ground and which is not short of water. The ETx is usually expressed in millimetres per unit of time, e.g. mm/day, mm/month, or mm/season. It can be determined experimentally using an evaporation pan, or theoretically, using measured climatic data, e.g. the Blaney-Criddle method. Since this study is not about finding ETx, we shall assume a table of the ETx as below for our scheduling computations. ETx is calculated for each field so as to have accurate information. **Table 1** shows the average Daily ETr estimated for each site so as to model the soil water requirement algorithm.

4.3. Scheduling Irrigations

The equation used [13] to estimate the irrigation requirement (IR) per plant is:

$$\text{GIR} = \left[\left(\text{CA} \times \text{Plant Factor} \times \text{ETx} \right) - \text{ER} \right] \div \text{IE}$$

where

GIR = the gross irrigation requirement in mm/d;

CA = plant canopy area in square mm;

Plant Factor = 0.95 for tea (approximated and it may need to be researched);

ETx = reference ET;

ER = expected rainfall mm/d;

IE = irrigation efficiency (assume 90% or 0.90 for low-tech drip system). These are the ones used in Tanzania for irrigation of tea farms.

If the overall water requirement is met by utilization of rainfall during crop growing period, then the net irrigation requirement on a particular day shall be determined by using the following formula:

$$\text{IRx} = \text{MDx} - \text{MD}(x-1) - \text{ERx} + \text{ETx} \quad \text{(refer above equations)}$$

Table 1. Average daily ETr (inch/day) estimates for different sites.

	Month					
Sites	April	June	July	Aug	Sept	October
Field 1	0.45	0.42	0.39	0.38	0.35	0.32
Field 2	0.41	0.45	0.42	0.39	0.30	0.28
Field 3	0.43	0.42	0.40	0.39	0.38	0.35

4.4. Database Analysis and Design

[10] have highlighted the building up of databases within a GIS for efficient water management. A coordinated approach at different levels of the irrigation system is required for improving irrigation management. The estimation of irrigation delivery, its schedule and duration is a key element in any irrigation system. This decision-making process is referred to as irrigation scheduling, the use of water management strategies to prevent over-application of water while minimizing yield loss due to water shortage or drought stress.

The GIS database is constituted by point, line and polygon themes data. Weather data refers to one or more years. Weather information is quite necessary for decision making. Accumulation of the weather data series are used, multiple simulations are performed to determine the frequency of crop water and irrigation requirements, or to perform an irrigation planning analysis relative to selected years such as dry, average and wet seasons.

The data should be stored in ASCII files format according to database requirements:

1) The moisture estimation table, including: MDx is moisture content depth in the field on the xth day, cm; IRx is amount of irrigation water supplied on the xth day, cm; ERx is rainfall received on the xth day, cm; x is irrigation period, day. This data is stored according to specific-areas(polygon coverage) of the farm
2) The farm fields (sensorLoc) table, containing the spatial and non-spatial information of the field that includes sensor ID and coordinates.
3) The evapotranspiration table, containing the sensor ID, evapotranspiration month and value ET.
4) The scheduling table, including the GIR = the gross irrigation requirement in mm/d, CA = plant canopy area in square mm, ER = expected rainfall mm/d and date calculated

4.5. General Architecture of the System

[6] suggested that in 3-tier architecture of GIS-based systems, the functional part of architecture includes a mapping service which produces and displays maps to the users. The mapping service interacts with the database service in order to retrieve data from the database, as required. File and communications services are only required in particular scenarios. The file service writes data into XML files, while the communications service interacts with other remote applications (like sensors) in order to receive live feeds of data, as required. In this research, user interface was used to interact with the mapping tool and application tool that collects data from the field and all data stored into the database. The mapping tool query the information from the database and present it into a user interface (web based mapping tool). **Figure 2** shows the relationships between different units of the system. User interface (front end tier) provides interactive tool that a user can communicate or request or query information from the system. There are middle tier applications that collect information from the farm or interact with Application Programming Interface (API) to get instant information of the farm. It is integrated with the application that can interpret geo-referenced data received from the farm. This application communicates with Google maps to provide state of art interface that can be interpreted by a farmer. For storing data, the system uses mySQL database to store all the collected field information.

4.6. Development of the Irrigation Scheduling Software

[5] stated that a program using GIS was developed for the spatial and temporal distribution of irrigation supply for a large-scale rice irrigation project in Malaysia. Their study focused on determining water deliveries on a periodic basis during the main (wet) and off (dry) seasons based on spatial and temporal demand. A water balance equation was used to determine the release for each constant head orifice (CHO) within the blocks and compartments.

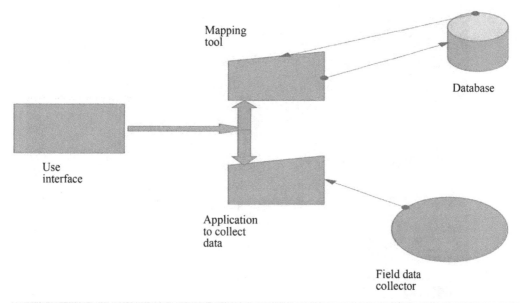

Figure 2. Three-tier web-based GIS applications architecture.

A user interface allows for the selection of a specific area and input of relevant information. The recommended discharges are displayed allowing the manager to view maps, tables and graphs. These provided a basis for decision making as the season progresses.

[6] states that the Google Maps API for Flash and the Flex API for ArcGIS Server allow users to use flex to embed the functionality of Google Maps or ArcGIS maps into web applications. Both APIs manage complex connectivity using the GIS package. They provided an easy to use interface for developers in order that Flash can be used to develop more interactive and rich Internet applications by users who do not need to understand Flash Action Scripts. This opens up a new way of developing GIS applications whereby the GIS server produces the map and then, Flash is used to add some more information to it. This information might entail adding a point, line or polygon, or equally might entail adding some more complex graphics to the map, e.g. custom navigation Flash controls.

In this study, decision support software system was designed and developed using Visual Basic (VB.NET) environment as a user interface platform and Google maps APIs as a GIS component. For creating treatment maps, then a popular open source class gmap.NET is used. Gmap.NET was great and powerful, free, cross platform, open source. NET control enable use routing, geocoding, directions and maps from Google, Yahoo!, Bing, Open Street Map, Arc GIS, Pergo, Sig Pac, Yandex, Mapy.cz, Maps.lt, iKarte.lv, Near Map, Ovi Map, Cloud Made, Wiki Mapia, MapQuest in Windows Forms and Presentation. It supports caching and runs on windows mobile. Hence, we developed functions that are useful for irrigation practices such as view maps, tables and graphs, updates, visualization, reports and irrigation schedule.
(http://greatmaps.codeplex.com/).

4.7. Google Online Maps

The software needs online maps from Google maps APIs. The system guided the users in seeing the farm and its' field sensors location online. The system can be integrated with switches to allow control of irrigation distantly. The maps display the fields and sensors allocation that is important to the farm manager. Many decisions in farm depend on spatial information from the day of planting to the day of harvest. Analysis like Yield maps can be produced by such maps. **Figure 3** depicts the software showing how sensors are allocated at the farms. The system has been customized to view the Ngwazi Estates in Tanzania. **Figure 3** shows the location of Ngwazi estate in Tanzania.

The system directly views the estate using Google maps using geo-reference information of the Ngwazi Tea estates. The center of the Ngwazi estate is estimated at the coordinate (−8.526253, 35.175208). **Figure 3** shows

Figure 3. Ngwazi tea estate (portion of the farm).

the map based information from the system. **Figure 3** shows points on the middle of each plot/field. There are 10 mid-points of the field. Each field is approximately 1 acre of the farm. Sensor location is shown by the pointers.

4.8. Moisture Estimation

The system developed provides a dialog box for water estimation. Information gathered from the sensors (rainfall, irrigation and moisture content) is crucial for estimation of the water that is lost due to several conditions like evapo-transpiration, run off etc.

In this system (**Figure 4**), the information is entered manually, but the effective way of entering data should be automatically or using a text file created by the sensor microcontroller or using API to query data from external data suppliers like satellite stations. Using the window below then we can enter the information obtained from the sensors. The user can click moisture Est button, the window (**Figure 5**) pops to receive moisture information of the specific sensor. For example, the sensor ID presents field sensor number. Each field has soil moisture sensor and the information can entered using the window manually.

After that, the user can be able to view the information using the system Geo-referencing window. By hovering the mouse then the information will be displayed instantly. The picture below shows the information from the sensor number at the center of the field number 8 (**Figure 4**). Also, using the export data button (**Figure 4**) then we can export the information easily.

5. Results and Discussions

5.1. Scheduling System and Formulas Used

The algorithm that has been used in the developed system is simple hence, more tests are needed to prove that it is efficient to use in intended fields. Nevertheless, the formulas have proven superiority in estimation of the Irrigation water. In scheduling, there is a button to click and choose and fill the appropriate fields and execute schedule. Go and see the results that have been created by the system. This spatial information will guide the farm manager to know exactly the need of the farm for the next day. Actually, knowing water required for irrigation is so crucial for farms that are in Tanzania due to drought conditions. Despite of Lake Ngwazi being near the farms sometimes water becomes the problem due to drought. The window below allows the prediction of the next day. The estimation needs the expected rainfall in hand, as the system cannot schedule without forecast

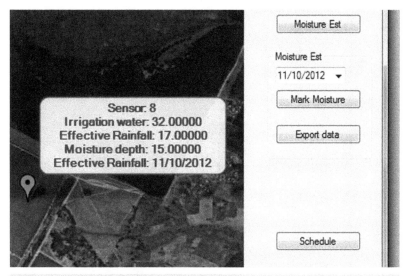

Figure 4. Georeferenced irrigation information.

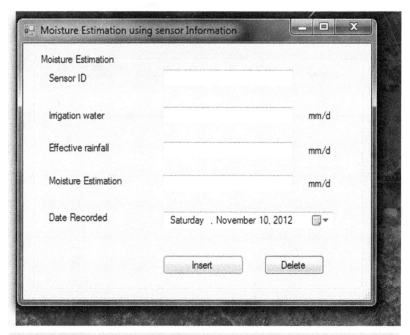

Figure 5. Moisture estimation window.

information of the rainfall. By clicking Schedule button as shown in **Figure 4** then scheduling window (**Figure 6**) pops to provide the information and schedule irrigation for tomorrow.

The scheduling window schedules the data for the next day. In fact, we expect this process to be automated if it is using microcontroller controlled system. When a user chooses Schedule for tomorrow then the system computes Gross irrigation requirement in mm/d and then checks if it's positive or negative. Negative will mean that the rainfall is very high for tomorrow hence there is no need to schedule for Irrigation.

5.2. Post-Processing of the Exported Information from the System

The system is able to export information (**Figure 7**) from the database hence extra processing with the exported data is possible. The system can export moisture information using the export data button.

Then, the user presses export data to excel button. This allows the user to have an exported Excel file that can be manipulated using ARCGIS.

Figure 6. Scheduling window.

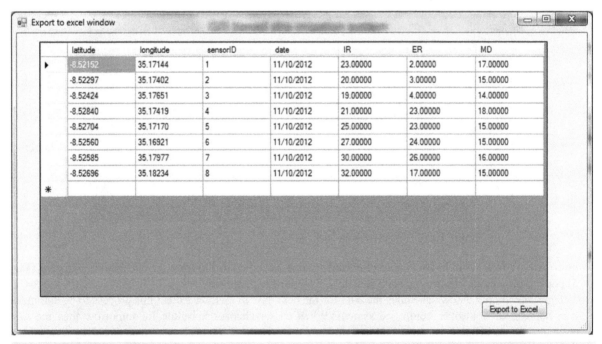

Figure 7. Export data window.

5.3. Analysis with ARCGIS

The exported data can be interpolated and analysed to see the distribution of the irrigation water and rainfall. In general, the expectation is a place with the lowest rainfall have an excess irrigation water while vice versa is also true. Therefore, this information related with Moisture depth of the soil so that other soil factors that influence Plant growth can be predicted. Consider the layout view of the irrigation water and rainfall below respectively.

Looking at both maps (**Figure 8** and **Figure 9**), we can realize that contradiction occurs at south east of the map were more irrigation more rainfall. This area indicates special site-specific characteristics. The soil of this

Irrigation Water distribution in Ngwazi tea estates

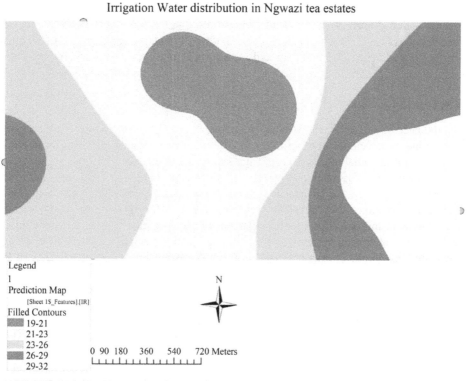

Figure 8. Irrigation water distribution.

Rainfall distribution in Ngwazi Teai Estates

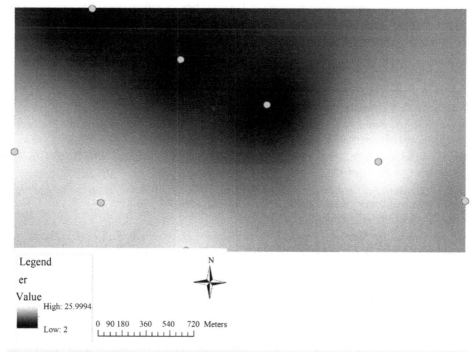

Figure 9. Rainfall distribution in Ngwazi tea estates.

place cannot hold water hence it needs more irrigation in high rainfall conditions. That means, it will consume more irrigation water. This place will need more sensors to measure moisture in small areas. This indicates the water table might be far away deep. In this sense, water in these areas will often be monitored so that the soil stays wet for acceptable crop production.

Further at the centre (**Figure 9**), while heading north, the site has low rainfall and low irrigation too. In fact this indicates that the place is so wet all season due to high water holding capacity of its soil. Such soil can retain water for further period hence its scheduling and monitoring might be loose as no any significant effect to low irrigation.

6. Conclusions

It's true that the necessity of achieving sustainable management of available water resources for irrigation supplies will determine the development of up-to-date and competitive agriculture. The information from developed system in this study is essential to assess crop water status and to efficiently irrigate tea crop as well as for improving water management. GIS with map-based user-interface technique linked with water management model can greatly assist to improve water management based on feedback from field information. This study presents comprehensible results along with new data sets and can assist irrigation managers to improve the decision-making process in the operation and management of the irrigation system. The system can improve the management of water allocation systems and scheduling water distribution system in existing schemes. This study has indicated that improvements in irrigation system management based on feedback of field information can satisfy the role of the precision agriculture. The system developed from this research actually provides an alternative approach to desktop way of evaluating data. Remote scheduling provides distant control over the conventional method where in field presence is required for monitoring of the crop. This gives managers an opportunity to get crop water conditions of the field while distantly located compared to the existing desktop monitoring where data need to be sent through email or whatever inconvenient existing methods. Interactive graphs provide great opportunity to monitor water.

In order to reduce costs, software developed from this study is necessary such that precision farming can be practiced to reduce wastes, minimize costs and improve crop yield and hence, increase profit in the same farm.

References

[1]   Official Online Gateway of the United Republic of Tanzania. http://www.tanzania.go.tz/agriculture.html

[2]   Kingalu, J. (2008) Drip Irrigation and Fertigation of Tea. Developing Agricultural and Agri-Business Innovation in Africa. http://info.worldbank.org/etools/docs/library/243684/session2aTzCaseStudiesTeaIrrigationMFPs.pdf)

[3]   Montgomery, G.E. and Schuch, H.C. (1993) GIS Data Conversion. *GIS Data Conversion Handbook*, 27-45. http://dx.doi.org/10.1002/9780470173244.ch2

[4]   Ozdilek, O. and Seker, D.Z. (2004) A Web-Based Application for Real-Time GIS. http://www.isprs.org/proceedings/XXXV/congress/yf/papers/934.pdf

[5]   Rowshon, M.K. and Amin, M.S.M. (2010) GIS-Based Irrigation Water Management for Precision Farming of Rice. *International Journal of Agriculture and Biological Engineering*, **3**, 27.

[6]   Adnan, M., Singleton, A.D. and Longley, P.A. (2010) Developing Efficient Web-Based GIS Applications. *CASA Working Papers Series*, Paper 153.

[7]   Fortes, P.S., Platonov, A.E. and Pereira, L.S. (2005) GISAREG: A GIS-Based Irrigation Scheduling Simulation Model to Support Improved Water Use. *Agricultural Water Management*, **77**, 159-179. http://dx.doi.org/10.1016/j.agwat.2004.09.042

[8]   Zairi, A., El Amami, H., Slatni, A., Pereira, L.S., Rodrigues, P.N. and Machado, T. (2003) Coping with Drought: Deficit Irrigation Strategies for Cereals and Field Horticultural Crops in Central Tunisia. In: *Tools for Drought Mitigation in Mediterranean Regions*, Springer, Netherlands, 181-201. http://dx.doi.org/10.1007/978-94-010-0129-8_11

[9]   Pereira, L.S., Teodoro, P.R., Rodrigues, P.N. and Teixeira, J.L. (2003) Irrigation Scheduling Simulation: The Model ISAREG. In: Rossi, G., Cancelliere, A., Pereira, L.S., Oweis, T., Shatanawi, M. and Zairi, A., Eds., *Tools for Drought Mitigation in Mediterranean Regions*, Kluwer, Dordrecht, 161-180. http://dx.doi.org/10.1007/978-94-010-0129-8_10

[10]   Rowshon, M.K., Kwok, C.Y. and Lee, T.S. (2003) GIS-Based Scheduling and Monitoring of Irrigation Delivery for Rice Irrigation System: Part I. Scheduling. *Agricultural Water Management*, **62**, 105-116.

http://dx.doi.org/10.1016/S0378-3774(03)00092-1

[11] Ortega, J.F., De Juan, J.A. and Tarjuelo, J.M. (2005) Improving Water Management: The Irrigation Advisory Service of Castilla-La Mancha (Spain).*Agricultural Water Management*, **77**, 37-58. http://dx.doi.org/10.1016/j.agwat.2004.09.028

[12] Teixeira, J.L. and Pereira, L.S. (1992) ISAREG, An Irrigation Scheduling Model. *ICID Bulletin*, **41**, 29-48.

[13] Allen, R.G., Pereira, L.S., Dirk, R. and Martin, S. (1998) Crop Evapotranspiration: Guidelines for Computing Crop Water Requirements. FAO Irrigation and Drainage, Rome, Paper-No 56.

6

Estimation of Soil Erosion Risk Using the Universal Soil Loss Equation (USLE) and Geo-Information Technology in Oued El Makhazine Watershed, Morocco

Asma Belasri, Abdellah Lakhouili

Faculty of Science and Technology, Hassan 1 University, Settat, Morocco
Email: belasri.asma@gmail.com

Abstract

Soil erosion by water is one of the major threats to soils in the north of Morocco; soil erosion not only decreases agricultural productivity, but also reduces the water availability. In the current study, Oued El Makhazine watershed is selected to estimate annual soil loss using the Universal Soil Loss Equation (USLE), remote sensing (RS) and geographic information system (GIS). GIS data layers including, rainfall erosivity (R), soil erodibility (K), slope length and steepness (LS), cover management (C) and conservation practice (P) factors are computed to determine their effects on average annual soil loss in the area. The resultant map of annual soil erosion shows a maximum soil loss of 735 t·h^{-1}·y^{-1}, about 65.25% (1575 km^2), of the watershed ranges between 0 and 95 t·h^{-1}·y^{-1}. Higher soil losses are observed at higher LS factor area. The spatial erosion maps generate with USLE method, remote sensing and GIS can serve as effective inputs in deriving strategies for land planning and management in the environmentally sensitive mountainous areas.

Keywords

Oued El Makhazine Watershed, Erosion, USLE, Geographic Information System

1. Introduction

Soil erosion represents one of the most serious land degradation problems [1]. It is defined as the loosening, dissolving and removal of rock materials from all parts of the earth's surface triggered by a complex interaction process of many factors: natural (climate, topography, soil, vegetation) and anthropogenic (tillage systems, soil conservation measures, overgrazing and deforestation) [2] [3].

In the north of Morocco, soil erosion affects negatively agricultural productivity, reduces water infiltration, underground water resources and water availability.

In Morocco, 40% of land is affected by water erosion [4]. In some parts of the Rif in northern Morocco, erosion rates sometimes reach 30 to 60 t·ha^{-1}·y^{-1} [5] [6]. Therefore, dams lose their water capacity initial storage due to their siltation which is estimated to 0.5% by year [7]. The largest Moroccan dams receive each year approximately 50 million tones of sediment [8], which affects their storage capacity and brings about an annual loss of almost 300 million Dirhams [9].

Due to the intensification of agricultural practices leading to unsustainable farming practices (e.g. inappropriate tillage practices, straw exportation, overgrazing) and specific bioclimatic conditions (e.g. recurring and severe droughts), more than 15 million hectares of the Moroccan agricultural land is under serious threat. As reported by Namr and Mrabet [10], it was estimated that out of 22.7 million hectares potentially exploitable in the Northern part environmental problem worldwide of Morocco, 77% were exposed to very high erosion risks [11]. The Global Assessment of Human Induced Soil Degradation (GLASOD) survey carried out during the 1980's by the United Nations Environment Programme (UNEP) and the International Soil Reference and Information Centre (ISRIC) established that the severity of human induced degradation had been classified as severe and very severe for more than 20% of the Moroccan territory [12].

In the Rif Chain (Northern Morocco), several local studies have been applied to evaluate and quantify the erosion risk, especially in the western and central Rif [13] [14].

Therefore, the estimation of erosion factors and exposed areas to soil erosion can be very helpful to identify the increment and the degree of the risks and, finally, to establish conservation measures and soil/water management plans.

In the current study, an effort to predict potential annual soil losses in Oued El Makhazine watershed in Morocco has been conducted using the Universal Soil Loss Equation (USLE) adopted in a GIS environment. It is used for the prediction of sheet erosion depending on the distribution of the aggressiveness of rainfall, the erodibility of soil, topography, land use practices and crop management.

2. Study Area

Located in the north of morocco, Oued El Makhazine watershed (**Figure 1**) covers an area of 2414 Km^2, and stretches from north latitude 34°45'15.3" to 35°15'44.6" west longitude 5°14'32.4" to 5°51'9.1", its altitude is between 10 m and 1677 m.

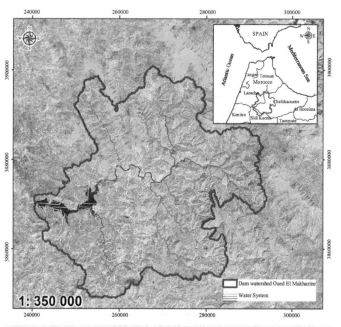

Figure 1. Oued El Makhazine watershed location map.

The study area consists west of plains with a very pronounced topography, while to the east the relief becomes more hilly and mountainous. The altitude increases gradually eastward with the first reliefs of the Rif mountain range, this configuration of the relief associated with the presence of the Atlantic Ocean to the west is the main cause of significant rainfall measured in the watershed, the study area receives an annual average rainfall of 1100 mm [15].

The predominance of impermeable soils in the watershed promotes runoff, which is increased by the slope effect as one progresses eastward.

The vegetation cover of the watershed is defined by the presence of matorral, typical Mediterranean landscapes, with a predominance of forests. Most of the forest cover is at the upper watershed. Downstream, the agriculture is the dominant economic activity in the watershed consists of cultivated plains.

3. Methodology

3.1. Data Processing and USLE Factors Generation

In recent years, GIS and remote sensing are often used to assess and map water erosion effects. Truthfully, with these modern techniques, it is increasingly exposed the advantages of spatialization methods for assessing and mapping soil erosion over large areas and setting up scenarios for rehabilitation.

Many techniques and studies realized worldwide were done about the evaluation of soil loss. Most of them are using the Universal Soil Loss Equation (USLE) and its revised version (RUSLE) [16]. Others had modified part of the equation to adapt in every country's situation. In this study, we tried to promote the process using the potentials of GIS.

The USLE equation is a product of five input factors (**Figure 2**) in raster data format: soil erodibility; rainfall erosivity; slope length and steepness; cover management; and support practice. Each factor varies over time and space and depend on other input data. Therefore, soil erosion within each pixel was calculated using the Universal Soil Loss Equation (USLE).

The USLE equation is described as:

$$A \equiv R \times K \times LS \times C \times P \tag{1}$$

where A is the computed spatial average of soil loss over a period selected for R, usually on yearly basis (t·ha^{-1}·y^{-1}); R is the rainfall erosivity factor (MJ·mm·ha^{-1}·h^{-1}·y^{-1}); K is the soil erodibility factor (t·ha·h·ha^{-1}·MJ^{-1}·mm^{-1});

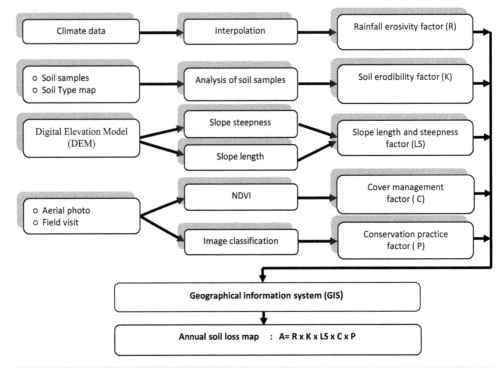

Figure 2. Flowchart of the methodology.

LS is the slope length steepness factor (dimensionless); *C* is the cover management factor (dimensionless, ranging between 0 and 1); and *P* is practices factor (dimensionless, ranging between 0 and 1).

In this study, the Universal Soil Loss Equation (USLE) was combined with GIS technologies to estimate the potential soil loss from areas within the Oued El Makhazine watershed, generate soil erosion severity maps, and analyze areas of critical soil erosion conditions which claim urgent need for convenient conservation measures and land management.

3.1.1. Rainfall Erosivity Factor (*R*)

The erosivity factor (*R*) in the USLE equation is accounted for by the rainfall-runoff, it is considered as a driver of soil erosion processes. The *R* factor represent the effect of raindrop impact and also shows the amount and rate of runoff associated with precipitation events, it is defined as the product of kinetic energy and the maximum 30 minute intensity and shows the erosivity of rainfall events [17].

Rainfall data of 10 years (2004-2014) was obtained from the Loukkos Hydraulic Basin Agency and imported into GIS environment since all the weather stations had geographical coordinates. Annual and monthly rainfall data of Oued El Makhazine watershed was used to calculate the *R*-factor in this study.

The rainfall erosivity values for the different stations were used to interpolate a rainfall erosivity surface using the Inverse Distance Weighted (IDW) interpolation method in ArcGis environment to generate a raster map for *R* factor. The IDW interpolation method was selected because rainfall erosivity sample points are weighted during interpolation such that the influence of rainfall erosivity is most significant at the measured point and decreases as distance increases away from the point. The IDW interpolation method is based on the assumption that the estimated value of a point is influenced more by nearby known points than those farther away [18] [19].

The Equation (2), below developed by Wischmeier and Smith [17] and modified by Arnoldus [20] was used in the computation:

$$R = \sum_{i=1}^{12} 1.735 \times 10 \left(1.5 \times \log_{10} \left(\frac{P_i^2}{P} \right) - 0.08188 \right) \tag{2}$$

where *R* is the rainfall erosivity factor (MJ·mm·ha^{-1}·h^{-1}·y^{-1}), P_i is the monthly rainfall (mm), and *P* is the annual rainfall (mm).

The mean values of *R* factor range from 74.63 MJ·mm·ha^{-1}·h^{-1}·y^{-1} to 116.11 MJ·mm·ha^{-1}·h^{-1}·y^{-1} (**Figure 3(a)**), approximately 76.82% of the rainfall erosivity factor R ranges between 74.63 MJ·mm·ha^{-1}·h^{-1}·y^{-1} and 96.59 MJ·mm·ha^{-1}·h^{-1}·y^{-1}.

3.1.2. Soil Erodibility Factor (*K*)

The soil erodibility factor (*K*) relate to the average long-term soil and soil profile response to the erosive power associated with rainfall and runoff. It is also considered to represent the rate of soil loss per unit of rainfall erosion index for a specific soil. The USLE model utilizes the technique proposed by Wischmeier [21] to measure the *K* factor of a soil type.

The specific methodology concerns many properties of soil such as gain size (silt, clay and sand), organic matter content, soil structure and soil permeability [21] [22].

The soil samples collected from Oued El Makhazine watershed were analyzed in soil laboratory of National Institute of Agricultural Research of Settat. For each sample the particle size analysis was performed in five classes (sand, very fine sand, silt, very fine silt, and clay) using Robinson's pipette method [23]. Organic carbon was measured by the Walkley-Black wet dichromate oxidation [24] and converted to organic matter by multiplying it by 1.724. Soil permeability code was obtained from the National Soils Handbook No. 430 (USDA, 1983) [25] and Soil structure code was determined using Soil Textural Pyramid [26].

To calculate the soil erodibility, the formula of Wischmeier and Smith [17] was used.

$$K = \frac{2.173 \times \left(2.1 \times M^{1.14} \times \left(10^{-4} \right) \times \left(12 - a \right) + 3.25 \times \left(b - 2 \right) + 2.5 \times \left(C - 3 \right) \right)}{100} \tag{3}$$

where *K* is the soil erodibility factor (t·ha·h·ha^{-1}·MJ^{-1}·mm^{-1}), *M* is the particles percentage (% of very fine sand + % of silt x (100% clay)), *a* is the organic matter content (% C x 1.724), *b* is the soil structure and *c* is soil permeability.

The analysis showed that 72 % of samples had clay texture, 18% Sandy clay loam and 10% Clay loam texture.

Erodibility is low for clay rich soils with a low shrink swell capacity, as clay particles mass together into larger aggregates that resist detachment and transportation [27].

Organic matter content varies from the lowest of 1.37 % up to the highest of 4.57%. The permeability of Oued El Makhazine watershed is very low, a result that was observed in numerous publications [28]-[30] that the presence of clay decreases the level of soil permeability.

The result (**Figure 3(b)**) shows that Oued El Makhazine watershed has soil erodibility, ranges from 0.24 to 0.85, the watershed has a moderate to severe soil erodibility with 71.26% of the area ranges between 0.68 t·ha·h·ha^{-1}·MJ^{-1}·mm^{-1} and 0.85 t·ha·h·ha^{-1}·MJ^{-1}·mm^{-1}.

3.1.3. Slope Length and Steepness Factor (LS)

The L and S factors in USLE reflect the effect of topography on erosion. It has been demonstrated that increases in slope length and slope steepness can produce higher overland flow higher erosion [31]. Moreover, gross soil loss is considerably more sensitive to changes in slope steepness than to changes in slope length [32].

Therefore one of the key factors in soil loss is topography, especially when the ground slope exceeds a critical angle. Topography factor plays a major role in soil erosion since it dominates the surface run off rate.

The combined LS factor of the watershed was derived from Digital elevation model by means of ArcGis environment. The computation of LS requires factors such as flow accumulation and slope steepness.

The flow accumulation and slope steepness were computed from the DEM using ArcGIS Spatial analyst extension.

The Equation (4), was adopted to calculate LS factor as proposed by Moore and Burch [33] [34].

$$LS = \frac{\left(\text{Flowaccumulation} \times \text{Cellsize}\right)^{0.4}}{22.13} \times \frac{\left(\sin \text{slope}\right)^{1.3}}{0.0896} \tag{4}$$

where flow accumulation denotes the accumulated upslope contributing area for a given cell, LS = combined slope length and slope steepness factor, cell size = size of grid cell (for this study 30 m) and sin slope = slope degree value in sin.

The LS factor value in the study area varies from 0 to 22 (**Figure 4(a)**). The Majority of the study area (77.72 %) has LS value ranges between 0 and 6.29.

3.1.4. Cover Management Factor (C)

The coverage factor ranges from 0 (full soil coverage) to 1(no soil coverage) [17].

C factor represents the effect of soil disturbing activities, plants, crop sequence and productivity level, soil cover and subsurface bio-mass on soil erosion [35]. Natural vegetation plays a predominant role in reducing water erosion [36].

Vegetation indices such as the NDVI (Normalized Difference Vegetation Index) are quantitative measures, based on vegetation spectral properties that attempt to measure biomass or vegetative vigor [37]. As an indirect estimate of vegetative density, a Normalised Difference Vegetation Index (NDVI), which approximates chlorophyll density, was calculated for the study area using Landsat TM images with a spatial resolution of 30 meter.

Satellite imagery acquired during the rain season are more adapted for this application given that soil erosion is most active and vegetation cover is at its peak during this season.

$$NDVI = \frac{NIR - RED}{NIR + RED} \tag{5}$$

where *NIR* is light intensity in the near-infrared, and *RED* is light intensity in the red band.

This index is an indicator of the energy reflected by the Earth related to various cover type conditions. When the measured spectral response of the earth surface is very similar to both bands, the *NDVI* values will approach zero. A large difference between the two bands results in *NDVI* values at the extremes of the data range.

Photosynthetically active vegetation presents a high reflectance in the near IR portion of the spectrum (0.76 - 0.90 μm), in comparison with the visible portion Red band (0.63 - 0.69 μm); therefore, NDVI values for photosynthetically active vegetation will be positive. Areas with or without low vegetative cover (such as bare soil, urban areas), as well as inactive vegetation (unhealthy plants) will usually display NDVI values fluctuating between −0.1 and +0.1. Clouds and water bodies will give negative or zero values.

The following formula was used to generate a C factor surface (**Figure 4(b)**) from NDVI values [38]-[40]:

$$C = e^{-a\left(\frac{NDVI}{b-NDVI}\right)} \tag{6}$$

where a and b are unitless parameters that determine the shape of the curve relating to NDVI and C factor. According to Van der Knijff an a-value of 1 and a b-value of 2 seem to give reasonable results [38].

3.1.5. Conservation Practice Factor (*P*)

The P factor is the soil-loss ratio with a specific support practice to the corresponding soil loss with up and down slope tillage [41]. Usually, the conservation practice factor corrects the USLE estimation for management and tillage practices that protect the soil from erosion.

In the present study the P factor map was derived from the land use/land cover and support factors. The values of P factor ranges from 0.55 to 1 (**Figure 4(c)**), in which the high value is assigned to areas with no conservation practices; the minimum values correspond to built-up-land and plantation area with strip and contour cropping.

4. Results and Discussion

Average Annual Soil Loss

The data layers (maps) extracted for K, LS, R, C, and P factors of the USLE model were integrated using Equation (1) within the raster calculator option of the ArcGIS spatial analyst in order to quantify, evaluate, and generate the maps of soil erosion risk and severity for Oued El Makhazine watershed.

Generally, if the estimated (A) value is high, it means a higher rate of sediment yield, while a lower value denotes a lower rate of sediment yield [42].

The Oued El Makhazine watershed was classified into five soil erosion risk categories (**Figure 4(d)**). The area and proportion of soil erosion risk classes are illustrated in **Table 1**. The maximum and minimum losses are respectively about 0 $t \cdot ha^{-1} \cdot y^{-1}$ and 735 $t \cdot ha^{-1} \cdot y^{-1}$, about 65.26% (1575 km^2) of the watershed ranges between 0 $t \cdot ha^{-1} \cdot y^{-1}$ and 95 $t \cdot ha^{-1} \cdot y^{-1}$.

The results shows that the Oued El Makhazine watershed is exposed to a very high erosion risk, the highest

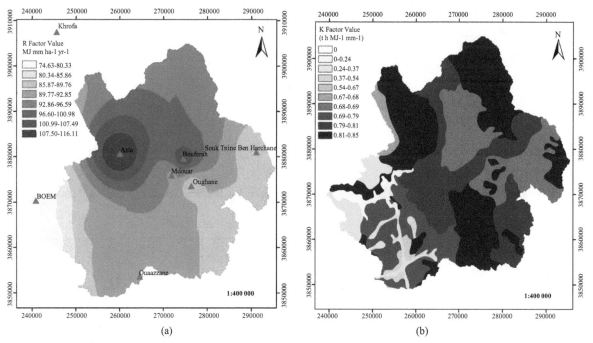

Figure 3. (a) Erosivity factor, (b) Soil erodibility factor.

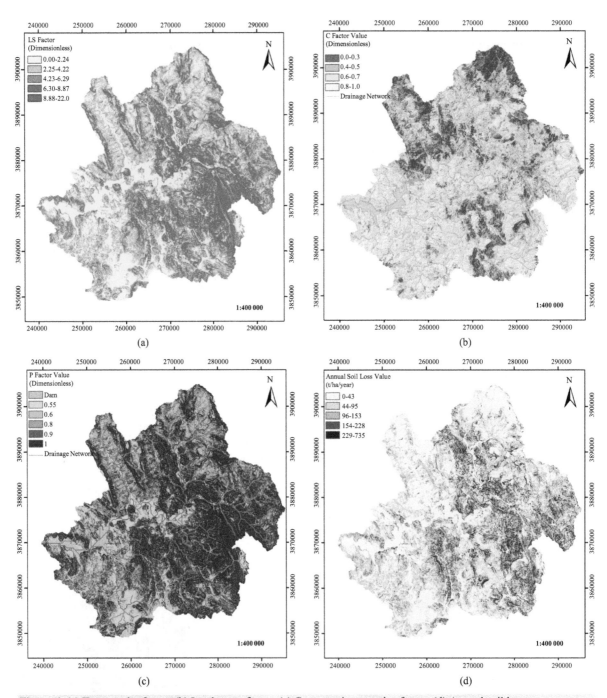

Figure 4. (a) Topography factor, (b) Land cover factor, (c) Conservation practice factor, (d) Annual soil loss.

Table 1. Classification of soil loss of Oued El Makhzine watershed according to area.

Soil Loss ($t \cdot ha^{-1} \cdot y^{-1}$)	Area (km^2)	Area (%)
0 - 43	798.81	33.09
44 - 95	776.47	32.17
96 - 153	491.74	20.37
154 - 228	263.58	10.92
229 - 735	83.40	3.45
Total	2414	100

soil loss values are spatially correlated with the steepest slopes.

5. Conclusions

USLE is a straightforward and empirically based model that has the ability to predict long term average annual rate of soil erosion on slopes using data on rainfall pattern, soil type, topography, crop system and management practices. In the present research, annual soil erosion rate map is generated for Oued El Makhazine watershed, a mountainous area, which represents the first reliefs of the Rif mountain chain. Several data sources are used for the generation of USLE model input factors and are stored as raster GIS layers in ArcGis environment.

The combination of the model USLE with GIS techniques has shown divers advantages, specifically the resulats of each factor involved in the erosion process. GIS allows rational management of a multitude of data, with respect to the various factors responsible of land degradation; in fact, in this case, it allows us to conclude that the primary factor responsible of the degradation of Oued El makhazine watershed is the steep and the slope morphologies, the soil erodibility and the vegetation cover comes in the second place. GIS also simplifies and facilitates the enrichment continuous updating of the database.

The intensiveness of the rainfall coupled with steep gradient slopes causes severe erosion runoff in the study area. The result of this high runoff and soil detachment is considered as the main agent for the high rate of soil erosion at Oued El Makhzine watershed.

There is an urgency to take into consideration this soil loss by all possible ways so as to reduce the existing amount of soil loss and to raise watershed rehabilitation and productivity.

Finally, the present investigation has demonstrated that GIS techniques are low cost tools and simple for modeling and mapping soil erosion, with the purpose of evaluating erosion potential risk for Oued El Makhazine watershed.

References

[1] Aiello, A., Adamo, M. and Canora, F. (2015) Remote Sensing and GIS to Assess Soil Erosion with RUSLE3D and USPED at River Basin Scale in Southern Italy. *Catena*, **131**, 174-158. http://dx.doi.org/10.1016/j.catena.2015.04.003

[2] Foster, G.R. and Meyer, L.D. (1972) A Closed-Form Soil Erosion Equation for Upland Areas In: Shen, H.W., Ed., *Proceeding of Sedimentation Symposium to Honor Prof. H. A. Einstein*, Vol. 12, Colorado State University, Fort Collins, 1-19.

[3] Kuznetsov, S., Feudel, U. and Pikovsky, A. (1998) Renormalization Group for Scaling at the Torus-Doubling Terminal Point. *Physical Review*, **57**, 1585-1590. http://dx.doi.org/10.1103/physreve.57.1585

[4] Chevalier, J.J., Pouliot, J., Thomson, K. and Boussema, M.R. (1995) Systèmes d'Aide à la Planification Pour la Conservation des Eaux et des Sols (Tunisie). Systèmes d'Information Géographique Utilisant les Données de Télédétection. *Actes du Colloque Scientifique International*, Hammamet, Tunisie, 4-12.

[5] Lahlou, A. (1977) Specific Degradation of Watershed in Morocco. Report n° 1000, Ministry of Equipment and National Promotion, Water Direction, Exploitation Division, Water Management Service, Rabat, 1977.

[6] Ait Fora, A. (1995) Modélisation Spatiale de l'Erosion Hydrique dans un Bassin Versant du Rif Marocain: Validation de l'Approche Géomatique par la Sédimentologie, les Traceurs Radio-Actifs et la Susceptibilité Magnétique des Sédiments. PhD Thesis, Sherbrooke University, Quebec.

[7] Tahri, M., Merzouk, A., Lamb, H.F. and Maxted, R.W. (1993) Etude de l'Erosion Hydrique dans le Plateau d'Imelchil dans le Haut Atlas Central. Utilisation d'un SIG. *Geo Observateur*, **3**, 51-60.

[8] Merzouki, T. (1992) Diagnostic de l'envasement des grands barrages marocains. *Revue Marocaine du Génie Civil*, **38**, 46-50.

[9] High Commission for Water, Forest and Combating Desertification (1996) National Watershed Management Plan.

[10] Namr, K.L. and Mrabet, R. (2004) Influence of Agricultural Management on Chemical Quality of a Clay Soil of Semi-Arid Morocco. *Journal of African Earth Sciences*, **39**, 485-489. http://dx.doi.org/10.1016/j.jafrearsci.2004.07.016

[11] Belkheri, A. (1988) Conséquences de la dégradation des bassins versants sur les retenues de barrages. Rapport du Séminaire National sur l'aménagement des bassins versants "Diagnostic de la situation actuelle", MARA., T.P., PNUD/FAO, 18-28 Janvier, Rabat, 23.

[12] Food and Agriculture Organization of the United Nations (2005) National Soil Degradation Maps. Soil Degradation Map of Slovenia. FAO/AGL, GLASOD, Last Update.

[13] Heusch, B. (1970) Prerif Erosion. A Quantitative Study of Water Erosion in the Marly Hills of the Western Prerif. Di-

rectory of Forest Research of Morocco, 129-176.

[14] Sadiki, A., Bouhlassa, S., Auajjar, J., Faleh, A. and Macaire, J.J. (2004) Use of GIS for the Evaluation and Mapping of Erosion Risk by the Universal Soil Loss Equation in the Eastern Rif (Morocco): Boussouab Watershed Case Study. *Scientific Institute Bulletin, Rabat, Earth Sciences Series*, **26**, 69-79.

[15] Wahby, Y. (2008). Modeling the Water Resources Management Model MIKE BASIN and Development of a GIS in the Loukkos Basin. Draft Report of Graduates, Mohammed V University, Mohammadia School of Engineers, Rabat.

[16] Renard, K.G., Foster, G.R., Weesies, G.A., McCool, D.K. and Yoder, D.C. (1997) Predicting Soil Erosion by Water: A Guide to Conservation Planning with the Revised Universal Soil Loss Equation (RUSLE). USDA Agricultural Handbook No. 703, 126-131.

[17] Wischmeier, W.H. and Smith, D.D. (1978) Predicting Rainfall Erosion Losses. A Guide to Conservation Planning. The USDA Agricultural Handbook No. 537, Maryland.

[18] Weber, D. and Englund, E. (1992) Evaluation and Comparison of Spatial Interpolators. *Mathematical Geology*, **24**, 381-391. http://dx.doi.org/10.1007/BF00891270

[19] Weber, D. and Englund, E. (1994) Evaluation and Comparison of Spatial Interpolators II. *Mathematical Geology*, **26**, 589-603. http://dx.doi.org/10.1007/BF02089243

[20] Arnoldus, H.M.J. (1980) An Approximation of the Rainfall Factor in the Universal Soil Loss Equation. In: De Boodt, M. and Gabriels, D., Eds., *Assessment of Erosion*, John Wiley and Sons, New York, 127-132.

[21] Wischmeier, W.H., Johnson, C.B. and Cross, B.V. (1971) A Soil Erodibility Nomograph for Farmland and Cons- truction Sites. *Journal of Soil and Water Conservation*, **26**, 189-193.

[22] Millward, A. and Mersey, J. (1999) Adapting the RUSLE to Model Soil Erosion Potential in a Mountainous Tropical Watershed. *Catena*, **38**, 109-129. http://dx.doi.org/10.1016/S0341-8162(99)00067-3

[23] SSEW (1982) Soil Survey Laboratory Methods. Technical Monographs No. 6, Harpenden.

[24] Nelson, D.W. and Sommer, L.E. (1982) Total Carbon, Organic Carbon, and Organic Matter. In: Analysis, A.L., Ed., *Methods of Soil*, 2nd Edition, ASA Monograph, American Society of Agronomy, Madison, 539-579.

[25] USDA (1983) National Soil Survey Handbook. No. 430, US Department of Agriculture, USDA, Washington DC.

[26] Hashim, G.M. and Wan Abdullah, W.Y. (2005) Prediction of Soil and Nutrient Losses in a Highland Catchment. *Water, Air and Soil Pollution: Focus*, **5**, 103-113. http://dx.doi.org/10.1007/s11267-005-7406-x

[27] ÓGeen, A.T., Elkins, R. and Lewis, D. (2006) Erodibility of Agricultural Soils with Examples in Lake and Mendocino Counties Oakland. University of California, Division of Agriculture and Natural Resources, Publication No. 8194.

[28] Bos, M.R.E. (1982) Prolific Dry Oil Production from Sands with Water Saturations in Excess of 50%: A Study of a Dual Porosity System. *The Log Analyst*, **23**, 17-23.

[29] Fuechtbauer, H. (1967) Influence of Different Types of Digenesis on Sandstone Porosity. *Proceedings of the 7th World Petroleum Congress*, **2**, 353-368.

[30] Herron, M.M. (1994) Estimating the Intrinsic Permeability of Classic Sediments from Geochemical Data. *Proceedings of the SPWLA 28th Annual Logging Symposium*, 29 June-2 July 1987, London, 23.

[31] Haan, C.T., Barfield, B.J. and Hayes, J.C. (1994) Design Hydrology and Sedimentology for Small Catchments. Academic Press, San Diego, 588.

[32] McCool, D.K., Brown, L.C. and Foster, G.R. (1987) Revised Slope Steepness Factor for the Universal Soil Loss Equation. *Transactions of the American Society of Agricultural Engineers*, **30**, 1387-1396. http://dx.doi.org/10.13031/2013.30576

[33] Moore, I.D. and Burch, G.J. (1986) Physical Basis of the Length Slope Factor in the Universal Soil Loss Equation. *Soil Science Society of America*, **50**, 1294-1298. http://dx.doi.org/10.2136/sssaj1986.03615995005000050042x

[34] Moore, I.D. and Burch, G.J. (1986) Modeling Erosion and Deposition. Topographic Effects. *Transactions of American Society of Agriculture Engineering*, **29**, 1624-1630. http://dx.doi.org/10.13031/2013.30363

[35] Prasannakumar, V., Vijith, H. and Geetha, N. (2011) Estimation of Soil Erosion Risk within a Small Mountainous Sub-Watershed in Kerala, India, Using Revised Universal Soil Loss Equation (RUSLE) and Geoinformation Technology. *Geoscience Frontiers*, **3**, 209-215. http://dx.doi.org/10.1016/j.gsf.2011.11.003

[36] Kheir, R.B., Abdallah, C. and Khawlie, M. (2008) Assessing Soil Erosion in Mediterranean Karst Landscapes of Lebanon Using Remote Sensing and GIS. *Engineering Geology*, **99**, 239-254. http://dx.doi.org/10.1016/j.enggeo.2007.11.012

[37] Agapiou, A. and Hadjimitsis, D.G. (2011) Vegetation Indices and Field Spectroradiometric Measurements for Validation of Buried Architectural Remains: Verification under Area Surveyed with Geophysical Campaigns. *Journal of Applied Remote Sensing*, **5**, Article ID: 053554. http://dx.doi.org/10.1117/1.3645590

[38] Van der Knijff, J.M., Jones, R.J.A. and Montanarella, L. (2000) Soil Erosion Risk Assessment in Europe. EUR 19044 EN, Office for Official Publications of the European Communities, Luxembourg, 34.

[39] Van der Knijff, J.M., Jones, R.J.A. and Montanarella, L. (1999) Soil Erosion Risk in Italy. EUR19022 EN, Office for Official Publications of the European Communities, Luxembourg, 54.

[40] Van Leeuwen, W.J.D. and Sammons, G. (2004) Vegetation Dynamics and Soil Erosion Modeling Using Remotely Sensed Data (MODIS) and GIS. *Proceedings of the Tenth Biennial USDA Forest Service Remote Sensing Applications Conference*, Salt Lake City, 5-9 April 2004.

[41] Renard, K.G., Foster, G.R., Weesies, G.A., McCool, D.K. and Yoder, D.C. (1997) Predicting Soil Erosion by Water: A Guide to Conservation Planning with the Revised Universal Soil Loss Equation (RUSLE). Agriculture Handbook No. 703, US Department of Agriculture, Washington DC, 1-251.

[42] Prasannakumar, V., Vijith, H., Geetha, N. and Shiny, R. (2011) Regional Scale Erosion Assessment of a Sub-Tropical High-Land Segment in the Western Ghats of Kera, South India. *Water Resources Management*, **25**, 3715-3727. http://dx.doi.org/10.1007/s11269-011-9878-y

7

Web Development of Spatial Content Management System through the Use of Free and Open-Source Technologies. Case Study in Rural Areas

Ioannis Pispidikis, Efi Dimopoulou

School of Rural and Surveying Engineer, National Technical University of Athens, Athens, Greece
Email: pispidikisj@yahoo.gr, efi@survey.ntua.gr

Abstract

The rapid development of the technology of the web systems and GIS, in conjunction with the world economic crisis, formed the content for the promotion and development of free systems and open-source technologies. At the same time, the tendency toward the standardization of data, metadata and services, with the aim of creating common "Language" for the reading and the dissemination of information available, is a basic research area in the global scientific community in this field. The development of WebGIS systems, taking advantage of the free technology, also contributes to finding more economical solutions, where the use of such systems is more directly accessible. The aim of this research work is, through the analysis of technologies for the Internet, and also the architecture of the WebGIS systems, to investigate the possibilities and to develop the appropriate free technologies, so as to design and implement a spatial content management system for the web. Search with the use of the latter, is the best response to the needs and visualization application maps, with scope in rural areas. In addition, reference is made to the existing content management systems, which provide both processing spatial data, and easily create a WebGIS application.

Keywords

WebGIS, Open Source Technologies, Free Systems

1. Introduction

Geographical Information Systems (GIS) are information systems based on data management with spatial and descriptive information. They are designed to support the collection, management, processing, analysis, modeling and imaging data referred in space and change over time. The main role of systems is to offer users powerful tools for solving complex spatial problems [1]. The Internet, even though created for the needs of the USA army, now forms an integral part of society. The widespread access to the Internet and the interactive content of the World Wide Web (WWW), have made this service a powerful means of exchange and management of information. Many applications in various fields have restructured and developed via the Internet [2]. The growth of the Internet affects the GIS in three different areas, such as the access of GIS data, the dissemination of spatial information and the modeling and processing of GIS [3]. The access and transfer of geographical data over the Internet is the first step in implementing a truly useful GIS system, where users will be able to work dynamically with the geographical data, without the need to install any specialized software. Independence from the latter was achieved through the establishment of appropriate services (OGC services), provided by the map servers. In addition, the technology AJAX (Asynchronous JavaScript and XML), which was conceived by Jesse James Garret, has strengthened the capacity of online systems, significantly reducing the burden on the Server. A major advantage over the use of this technology is the fastest response of the interface, to provide for the possibility for real-time applications [4]. Therefore, the evolution of the Internet and the development of technologies that support the creation of online systems, have led to WebGIS applications where through standardization of data, metadata and services, the exchange and analysis of geographical information are easy, direct, economical and efficient. In addition, the need to create more economical solutions resulted in the development of open source technologies at all levels of architecture of a WebGIS system. In order to easily exploit these technologies that do not require specialized knowledge, Content Management Systems (CMS) were developed. Example systems with processing and analysis of spatial data are the OpenGeo Suite, which is a commercial product but implemented by free technologies, the Cartaro, which provides geospatial functions and Web services in CMS Drupal, the Map Guide OpenSource and the GeoNode.

Based on the fact that the basic architecture of a WebGIS system consists of three basic levels (Client-Server-Data Server), reference is made to the technologies used for the implementation of each one of them. In particular, the level of client utilized the AJAX technology, where through the JSON format data are received and sent to the Server, asynchronously in the background, without the need to renew the application [5]. In addition, it was based on the ExtJS Framework, considered to be the most suitable for the development of applications on the Internet [6]. Finally, web mapping capabilities were provided through the OpenLayers library. At the level of the server the web programming language PHP was used with web server, the Apache HTTP server and map server—the GeoServer. The latter is integrated with the auxiliary tool GeoWebCache, through which the response of the system becomes faster and more efficient and reduces the workload of the GeoServer [7]. At the level of the Data Server the database PostgreSQL was used and, specifically, the PostGIS, so as to have geographical capabilities [8].

The content management system that was created provides the ability to manage issues relating to processing user requests and maps. In addition, it supports the organization and transfer of spatial data in the database and the GeoServer, as well as the processing in real-time via WFS-T (Web Feature Service-Transaction) which supports the GeoServer. The aim of the system is to be used to resolve bureaucratic and time-consuming problems in supplying land distributions and land reform maps in rural areas. In addition, it covers the requirements of both the administrator and the user, while its installation is direct and easy to use.

2. Architecture of WebGIS

By the term architecture, we mean all mechanisms, libraries and software to be used in a geographical information system to be complete. A WebGIS is composed of three basic levels; the first level is the client, while the second level consists of a team of servers and software, operating on the same or different server. The software called server-side software actually activates the use of server-side web programming languages (Java, PHP, Python, etc.). As server group we mean the web server and map server, and finally, the third level consists of the data server, including the database.

2.1. Client

The client is a browser which has each user on the computer and is necessary for browsing the web, providing communication between the user and the web server. The communication with the user comprises the web page and asynchronous data exchange via the client-side web programming languages (JavaScript). When a user enters an address, the client receives the request and via the HTTP protocol sends it to the corresponding web server where this website is stored. The web server responds to the request by providing the web page requests.

For the asynchronous exchange of data between the client and the user, scripts are used written in the client-side web programming languages. The scripts are mostly functions of JavaScript, which are carried out and modify accordingly the web page with the implementation of an event. An example event is the click of a button (onclick), whose syntax is shown below.

```
<input name= "example" type= "button" onClick= "function()
" value= "event" />
```

2.1.1. AJAX (Asynchronous JavaScript and XML)

AJAX is used to describe modern technologies, techniques and methods, which are not necessarily related to the components of the AJAX (Asynchronous JavaScript and XML). The basic idea is to no longer need to renew the entire page, in order to send data to the server. In particular, the whole process is asynchronously processed in the background, while renewing only the part of the page that needs updating. The operation of the AJAX is based on the JavaScript XML Http Request [9].

Main advantage of using AJAX is that only the necessary data are sent and received and therefore the traffic of the server is reduced, since only part of the requested page is renewed. Finally, another advantage of its use is the fast response of the interface, as well as the possibility to provide online applications [4].

The exchange of data via AJAX is carried out using various formats. The most commonly used are summarized below:
- HTML

The most common format for the exchange of data via AJAX; it makes it easy and simple to update section of the page, by simply assigning the received data to the inner HTML attribute of a web page element.
- XML

It consist the main data-exchange format for which the techniques of AJAX were conceived.
- JSON (JavaScript Object Notation)

This format started to gain ground in recent years. Compared with the XML, for the same data, less space is required, while it is easier to parse the data.

More formats of data transfer are also used, such as the JavaScript and CSV (Comma Separated Value) [4].

2.1.2. DOM (Document Object Model)

The Dom is an independent platform and language convention, which provides a structured representation of the XML and (X)HTML document and specifies how this can be accessed from any programming language, so that it is possible to change the structure, the content and the style of the document. The structure of the document is represented as a tree composed of elements (element nodes), text (text node), properties (attribute node) and comments (comment nodes). The root of the tree is a document node, while the nodes are governed by hierarchy relations. The first element node of the tree is called root node and each element except the root has exactly one parent. The elements that have the same parent are siblings and descendants of the parent element. In the following hierarchy example (**Figure 1**), the head and body elements are siblings, descendants of the HTML, while their children are the elements with which are put together [10].

On the basis of the hierarchy of the DOM tree, the various scripts choose the elements on which they want to act. The selection methods are presented in **Table 1**.

Finally, a standard feature that is achieved through the DOM and in particular the document node, is the creation and deletion of data from the website, as well as the modification of the characteristics of existing data with the use of client-side programming languages, such as JavaScript [10].

2.2. Web Server and Application Server

The Web server is the software that responds to requests from the client via the HTTP protocol. It is designed to respond effectively to requests of a large number of clients and send static files. The request is sent via HTTP on

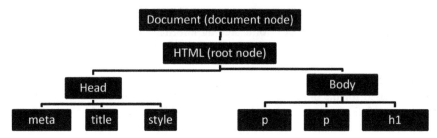

Figure 1. Example of DOM tree.

Table 1. Methods of data selection.

Selection Modes	Selection Method
Id	Document. Get Element by Id ("Contents")
Tag	Document. Get Element by Tag Name ("p")
Name	Document. Get Element by Name (Name)
Class	Document. Get Element by Class Name (Class Name)

the Web server, who is looking in the files of the hard drive of the server, if this file exists. A file will not be found in the server when not spelled correctly or when the request is not given with the correct path. If the Web browser is not configured to display a particular file type, then the user is asked to save the file on his local drive.

The most popular Web servers that are in use today are the Apache HTTP server and the Internet Information Services (IIS). The Apache runs on all modern operating systems, including Windows, Linux, Mac OS X and Unix. It is released under the license of Apache software, as open source software. It is serviced by an open source community and supervised by the Apache Software Foundation [11].

The weakness of the web server to manage and return only static documents comes to cover the Application server. It manages both the dynamic content of web pages through the supported script engine, and the simultaneous requests from users. Therefore, the Web server expects from the application server to return the result of the dynamic content and it in turn to do the job of returning the final static file to the client. Through the application server it becomes possible to communicate with the database and with other servers, such as the Map server.

2.3. Map Server

The Map server is a type Application server with manageability, processing and visualization of spatial data. The main feature of Map servers is the acquisition of spatial data from a spatial database and their dissemination on the Web, by using appropriate geospatial standards and services. A Map server may be Web server, so called web-mapping server or installed on a web server that has the appropriate server-side programming languages to support [7].

The most popular web-mapping servers used are the GeoServer, the MapServer and the ArcGIS server. The first two are open source web-mapping servers, while the third is commercial.

2.3.1. Web Map Service (WMS)

Through the WMS service, georeferenced images are available online only for viewing and not for further data processing. The client sends a request to the map server, and, based on the parameters of the request, the map server generates the final image and answers the request. The image produced by the WMS is the final product of the request and is provided by vector or normalized data or by a combination thereof.

The request for a WMS standard must be either in the form Get Capabilities or Get Map. The first one is used for the metadata of spatial data in XML format document. The second returns as result a georeferenced image. Such a request is considered as complete only when it includes information concerning the request, the name of the layer as recorded on the map server, the style of the layer, the reference system, the search limits and ultimately the size and format of the final image created.

The Get Feature Info is optional request for searching information on the elements of a map produced by the WMS service on the basis of the pixel (i, j) value of the image [12].

2.3.2. Web Feature Service (WFS)

The WFS service is used for the direct use of vector data, returning the actual geometry and characteristics of the latter. For the description of spatial data the language GML (Geographic Markup Language), which is an extension of XML, is used.

The main demand that contains a service WFS is in the form Get Capabilities, Describe Feature Type, Get Feature, Lock Feature and Transaction. Through the Get Capabilities the user receives the metadata of spatial data in XML document format.

```
Http://example.com/geoserver/wfs?
 service=wfs&
 version=1.1.0&
 request=GetCapabilities
```

With the Describe Feature Type appear in GML format, information on either a layer or for a specific element of the layer.

```
Http://example.com/geoserver/wfs?
 service=wfs&
 version=2.0.0&
 request=DescribeFeatureType&
 typenames=namespace:featuretype
```

With the Get Feature request return to the user the real vector data with their geometry and their descriptive features.

```
Http://example.com/geoserver/wfs?
 service=wfs&
 version=2.0.0&
 request=GetFeature&
 typenames=namespace:featuretype&
 featureID=feature
```

The above example is a request via HTTP for the feature of a specific Feature type layer, as created and registered into the GeoServer.

The Transaction request provides the ability to create, modify, and delete data which have been published through the WFS service. The WFS service which supports this request is called WFS-T (Web Feature Service-Transaction). Finally, the Lock Feature is used in order to protect the data from the request WFS-T [13].

2.3.3. Web Coverage Service (WCS)

The WCS concerns the WFS service for the normalized data (raster). It is a service which makes possible to access grid coverage's data online. The grid data refers to satellite imagery, digital aerial photographs, digital terrain models and phenomena that can be represented by values at each measuring point. The request forms supported by the WCS service are the Get Capabilities, the Describe Coverage and the Get Coverage. The first request returns the metadata of this spatial data. The Describe Coverage returns in XML document format the full description of coverage request. Finally, the Get Coverage returns the final request in image format.

It should be noted that there should be no confusion between the two services WMS and WCS. Via the WMS service the spatial data is visualized in raster format. The end result is an image, in which there is no possibility for further analysis and processing. On the other hand, with the WCS, the image is a grid coverage, which the user can access via the requests supported by this service [13].

2.3.4. Web Processing Service (WPS)

The WPS service provides GIS functions to clients on the web, including access to planned in advance calculations and computational models, which are associated with spatial data [14]. It can provide simple calculations, such as the definition of a buffer zone or complex, as a model of global climate change. The WPS aims to edit both vector and raster data. It should be noted that with this service successive algorithms and procedures apply on some data. Therefore, the results will assist decision-making in complex problems. The requests that may be created by the WPS service are the Get Capabilities, the Describe Process and the Execute. With the first request form, information about the WPS service is acquired in XML format. With the Describe Process, information about the procedure that has been selected to run is provided. In particular, the XML format contains information relating to the name of the process, the input spatial data, the parameters and the format of the data output. Finally, through the Execute request, the procedure provides the final result of the processing of the input spatial data [14].

2.3.5. Geography Markup Language (GML)

The GML is an extension of XML language and is defined in order to express and transfer geographical features on the Internet. It is an international standard that is able to integrate not only vector features but also rasterdata and data from sensors. The possibility of integrating of the latter, implemented with the GML version 3.0, is important for its usefulness.

The GML encodes the geometric objects as elements in a document. The types of geometric objects that support the versions GML 1.0 and GML 2.0 are points, lines and polygons. The GML version 3.0 incorporated new structures, which support coverage information, normalized structures and sensors' data.

The features of GML represent physical entities, which may contain geographical and/or descriptive data (e.g. rivers, bridges and buildings). Features describing the physical entity of an element should not to be confused with the geometric objects, since the latter define a location or area. Therefore, a feature may comprise one/ more or not any geometric objects. Such a physical entity that may contain more than two geometric objects is a building described by its location and surface area.

The geometric objects are identified by coordinates, which are recorded within the following tags:

```
<gml:coordinates panel> 30.45, 54.21 < /gml:coordinates panel>
<gml:pos> coordinates panel  < /gml:pos>
<gml:posList> coordinates panel  < /gml:posList>
```

With the use of the label <gml:coordinates>, as shown in the example above, it is not possible to access the coordinates separately (e.g. 30.45), given that the contents of this label is a string. From the GML version 3.0 and then, labels <gml:pos> and <gml:posList> are used for recording coordinates in order to allow accessibility of individual coordinates via the XML DOM.

The CRS determines the reference system of the coordinates of the geometric objects of a GML document.

```
<gml:LineString gml:id= "p21 "srsName= " / .. /EPSG/ 4326 ">
<gml:posList>45.67 88.56 55.56 89.44 < /gml:posList>
< /gml:LineString >
```

In accordance with the above, a line geometric object was created, of which the reference system is determined through srsName feature. The value of the latter is a URL of an XML document of the specific reference system based on the EPSG (European Petroleum Survey Group) codes. In particular, the code 4326 defines the WGS84, while the corresponding XML document can be accessed through the website [15].

2.3.6. KML Encoding Standard (KML)

The Keyhole Markup Language (KML) language is an XML which focuses on the visualization of geographic information, including notes on maps and images. The geographical visualization includes, not only the representation of spatial data on the globe, but also the control of the user's navigation. Initially developed to be used with the Keyhole Earth Viewer (by the Keyhole company, funded by CIA), which in 2004 was acquired by Google. The application has been further developed known as Google Earth, which is currently very popular, with millions of users worldwide. Given this popularity, the KML has been widely spread and has nowadays been incorporated in many geographical information management systems, as a means of interconnecting with Google Earth. In 2008 the KML has been approved as an international standard by the OGC. The KML is complementary to most standards of OGC including the GML (Geography Markup Language), the WFS (Web Feature Service) and the WMS (Web Map Service). The current version (2.2) of the KML incorporates geometry elements of the GML (v2.1.2). These include points, lines, line strings, linear rings and polygons.

The basic structure of a KML is based on the general principles of the XML (Extensible Markup Language) language, which uses tags in order to determine the beginning and the end of the different properties included in the file.

```
<kml xmlns= "http://www.opengis.net/kml/2.2">
<Document>
DEFINITION OF GENERAL PROPERTIES
<Placemark>
DEFINITION OF SPATIAL AND DESCRIPTIVE DATA
< /Placemark>
< /Document>
< /kml>
```

The label: <kml xmlns= "http://www.opengis.net/kml/2.2"> is placed at the beginning of the KML and indicates that the file will follow the rules of the KML (v2.2) standard of the OGC. All data, together with the definition of the general properties, are registered within the label <Document>.

Within the label <Placemark> the descriptive and spatial data of a geometric object are placed. The latter may be a point, a line or a polygon.

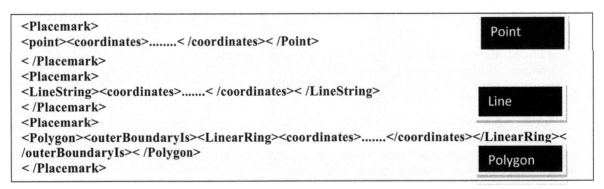

It should be noted that the coordinates of each geometric object are defined within the <coordinates> label [16].

2.3.7. GeoJSON

The GeoJSON [17] is an encoding format of geographical information, based on data transfer standard JSON (JavaScript Object Notation). Compared with the two previous forms, the GeoJSON is not considered by OGC as an international standard but as the result of a web developers' team. A GeoJSON may represent geometry, a feature or a collection of features. In a GeoJSON an object consists of a set of parts, which are described by a name and a value. The name is always a string, while the value may be string, number, object, table, true/false or null.

The geometrical objects which support a GeoJSON are points (Point), lines (Line String), polygons (Polygon), multiple points (multipoint), multiple lines (Multi Line String), multiple polygons (Multi Polygon) and collection of geometric objects (GEOMETRYCOLLECTION).

A feature in a GeoJSON contains a geometric object together with its properties. In particular, a feature is described by two pairs of names and values. The first is the geometry, which takes the elements of a geometric object and the second is the properties, where the feature attributes are recorded.

```
{ "type": "Feature",
  "geometry": { "type": "Point", "coordinates": (102.0, 0.5) },
  "properties": { "prop0 ": "value0 "} }
```

The reference system in a GeoJSON is determined through the "crs" object, which is described by two names. The first is the type where the EPSG value is assigned, while the second is the properties which consists the subject where various pairs of names and values are assigned. The most important is the code, which takes as value the corresponding EPSG code of the reference system. If the reference system is not defined, then the WGS 84 is considered as default. It should be noted that depending on the reference system, the description of the crs object can be found in the website:

```
"crs": {
   "type": "EPSG"
   "properties": {
     "code": 4326
   } }
```

http://spatialreference.org/.

2.4. Data Server

The Data server distributes the data in a database. Usually, in case of spatial data, the database connects with the map server and the query is executed by the server. For non-spatial data, the acquisition of data from the database can be accessed through the Application Server and in particular through the available server side programming language. For the database querying, whether spatial or non-spatial, the SQL (Structured Query Language) language is used. The most popular databases used in web applications are PostgreSQL, MySQL and Oracle. The first two are open source databases, and Oracle is a commercial product. MySQL is not suitable for spatial data, while PostgreSQL with the integration of PostGIS spatial database is considered as the most appropriate.

3. Design and System Function

Within the content of this research work, an online system was created, based on free systems and open-source technologies, consisting of three discrete subsystems. The first one is used for the easy and quick implementation of the necessary elements of the system. The second is a spatial content management system, and the third was created for web querying of information that relates to land consolidation or distribution work for a specific parcel.

The development of the application was based on the ExtJS Framework, considered to be the most suitable for the development of applications on the Internet [6]. Finally, web mapping capabilities were provided through the OpenLayers library. At the level of the server the web programming language PHP was used with web server the Apache HTTP server, map server the GeoServer and Database the Postgre SQL/PostGIS (**Figure 2**).

3.1. Methodology

The methodological approach adopted in order to implement the system consists of three stages. In the first stage the system's requirements were reported with regard to the users' management, the personal data, the GeoServer, the need for determining the competence of the application area and the data regarding the web request of the citizens. Then the appropriate tables were created in the spatial database PostGIS and exported into

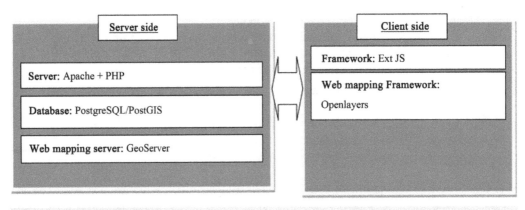

Figure 2. System architecture.

"sql" file. In addition, the most appropriate free technologies that can be utilized at the various levels of the architecture of the online system were investigated. As a result, the development of the application was based on ExtJS framework and therefore the basic language of implementation is the JavaScript. It should be noted that in this Framework the AJAX technology is integrated. PostgreSQL/PostGIS was used as database, while to get data from it, the server-side programming language PHP was used. For the publication of spatial data on the Internet the Web-mapping server GeoServer has been used, while their web cartographic representation was through the OpenLayers. Finally, the Apache HTTP server was used as server.

In the second stage, the web application of the system's implementation was created. The user fills up tabs with data relating to personal information and data concerning the connection to the spatial database. Then, by using this information and the "sql" file, the appropriate tables are created in the database; at the same time, a "php" file that contains the login information to the database is also created. In addition, implemented installation of GeoServer with the following plugin: geoserver-xx-SNAPSHOT-printing-plugin, in order to be able to print data in pdf.

In the third stage, the interface was created both for the spatial content management system and the web application for the online users' requests. The first application was created in a way to be connected with both the database implemented in the preceding step and the GeoServer.

The system thus created (**Figure 3**) offers the user the online data search concerning a land parcel. By implementing this application, an automated and friendly system is created both for the citizens and the surveyor engineers that request data on parcels that are included in a land consolidation or distribution project area. The WebGIS application consists of two parts; the right part shows is complemented by the citizen, while in the left part there is the geospatial map, with various tools for the correct completion of the form. At the top right of the window two more buttons provide communication and online help to the users.

The spatial content management system (**Figure 4**) consists of seven main sections for the users' management: the web-mapping server GeoServer, the online applications, the contact information and the processing of layers. Access to the system is provided via the username and the password created during the installation process.

3.2. Case Study

Poros of Evros Prefecture, located in the region of Eastern Macedonia-Thrace, was chosen as study area. Evros Prefecture is considered as a plain with only 10% mountainous, while 62% is covered by a fertile plain related to the Evros River with its streams and dense irrigation system. From the early 20th century, many land distribution and re-allotment projects took place in the study area, in order to deal with the problem of land fragmentation.

3.2.1. Materials

The choice of this particular region was based on the availability and accessibility of distribution and re-allotment maps to be used for the application procedure. The available data are in raster format, georeferenced in the Hellenic Geodetic Reference System HGRS87. The whole application process is based on raster data, although the spatial content management system created also deals with vector data.

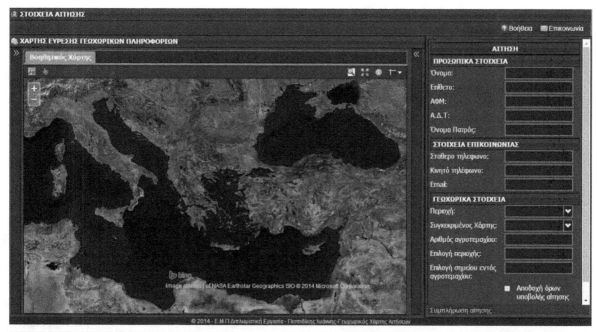

Figure 3. Web application for the online users' requests.

Figure 4. Display of the spatial content management system.

3.2.2. Management System for Data Preparation

Having implemented the application, the administrator of the system makes the necessary updates by defining the limits of access, by starting up the GeoServer and by uploading the available maps for viewing (**Figure 5**). New users that are in charge for the processing of the citizens' application request may be added.

3.2.3. Application of Maps' Request

At first, the citizen can search for a specific land parcel by using the map application, which has as background satellite imagery provided by Bing (**Figure 6**).

It should be noted that, if digitization of maps was carried out, then it would be easier for the user to search information concerning specific land parcels based on pre-set queries in the database. In particular, by selecting a point within the parcel, the spatial database would be automatically queried for information regarding the specific

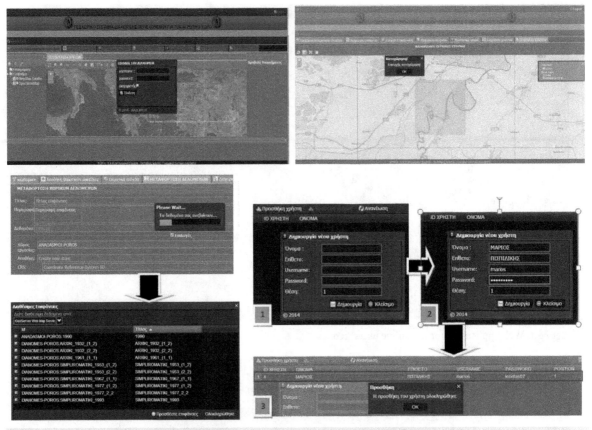

Figure 5. Initial update of the system (examples: login, specify area zone, data upload, creation of new user).

Figure 6. Example of a land parcel search.

parcel, thus, easing the citizen's search.

In this example the land parcel queried is located in an area of Poros where land consolidation occurred in 1990. The user by using tools (such as "transparent"), can easily detect the identification number of the land parcel which is 220. Then, the user completes the application. Personal information is filled up first along with contact details, and then, the geospatial data are added (**Figure 7**). After finalizing the completion of the application, the user has to accept the terms of validity required. Then, by activating the "Send request" button, the user can submit the final application. The completeness of the submitted application is checked, and the information is either transferred to the database or, in case of missing data, the user gets a relative notice.

3.2.4. Management System for Processing of Applications

When the user authorized by the administrator, chooses the "Management of applications" option of the CMS, the citizens' requests are displayed (**Figure 8**).

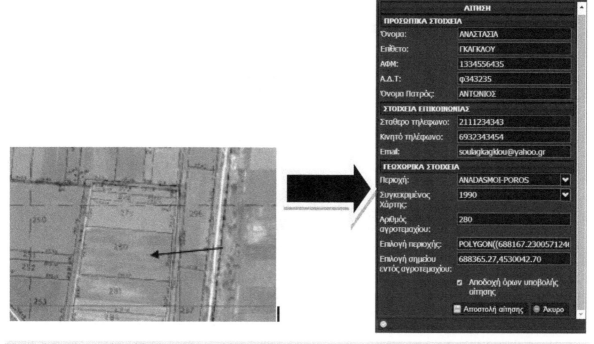

Figure 7. Land parcel identification and completion of the application form.

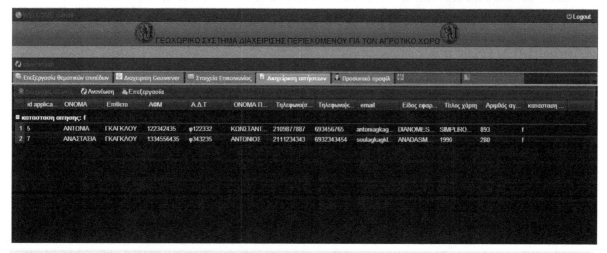

Figure 8. Applications available within the spatial content management system.

Then, by choosing one of these two requests and by selecting the "Edit" button, the following WebGIS application is showed.

The user can visually select the area of the map to be extracted and choose its resolution in dpi units (**Figure 9**). Finally, the user may select the "completion of application" in order to close the relative window and update the database as well as the system that the user carried out the specific processing task.

4. Conclusions

The design and implementation of this system were based on state-of-the-art technologies through which, the application was smoothly and fast operational at all levels. In particular, the AJAX technology was used, and through the JSON format data are downloaded from the Server asynchronously in the background without the need to renew the application. Additionally the system was based on the ExtJS Framework, considered as the

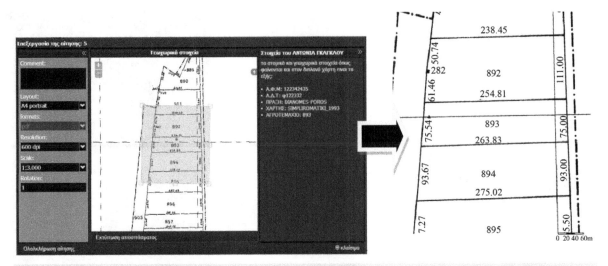

Figure 9. Generation of pdf with the extract of the selected area of the map.

most suitable for the development of web applications. The GXP library was also utilized, which includes a large number of cartographic tools for the Internet. By exploiting the GeoServer capabilities and the GeoWeb-Cache tool, the application became faster, since the response to the various spatial clients' requests was quick and efficient.

The implemented spatial content management system provides management capability in issues relating to the users' processing, the organization and transfer of spatial data in the GeoServer, as well as their online processing through the WFS-T supported by the GeoServer.

Some improvements that could be incorporated into the system in order to integrate it may include:

– Use of the GeoExplorer that provides additional possibilities to publish WebGIS applications according to users' needs.
– Further exploitation of GeoServer roles that are created to serve security functions. These roles are assigned to users/group and determine what functions are allowed in the GeoServer. Such roles are: the accessibility to various layers, the access to WFS services and the manageability of the GeoServer.
– Use of CSW (Catalog Service-Web) for database querying based on the metadata.
– Exploitation of the GeoServer capabilities, which is the available plugins, such as inspire, CSW, WPS and SQL server. The first two relate to the metadata. The WPS concerns the option of providing GIS functions including access to pre-calculated and computational models, which are associated with spatial data (raster and vector). The SQL server allows for the creation of SQL queries, based on pre-defined parameters for the specific spatial information.

Proposals for future development of the web management system described in this paper, both in theoretical and practical terms, may include:

– Further investigation of commercial and free spatial content management systems, as well as comparative survey investigation between them.
– Further investigation of commercial and free Map server with additional comparison between them.
– Economotechnical feasibility study for the creation of a web application to resolve a problem with commercial and free systems, including software cost, time and work required.
– Redesign of the database system, based on international standards, such as the LA.D.M. (Land Administration Domain Model).
– Knowledge exploitation, to investigate and implement web application for the Hellenic Cadastre.

References

[1]   Stefanakis, E. (2003) Supports Geographical Data and Geographical Information Systems. Papasotiriou Publication, Athens. (Greece)

[2]   Plewe, B. (1997) GIS Online: Information Retrieval Dialog Box, Mapping, and the Internet. Word Press, Santa Fe, New Mexico.

[3]   Sotiriou, I.E. (2010) Comparison and Integration of WebGIS Software. Diploma Thesis, Geoinformatics, NTUA.
      http://dspace.lib.ntua.gr/handle/123456789/3179

[4]   Athens University of Economics and Business (2011) AJAX: Applications on the Web.

[5]   Darie, C., Brinzarea, B., Cherecheş-Toşa, F. and Bucica, M. (2006) AJAX and PHP. Building Responsive Web Appli-
      cation. Packt Publishing Ltd., Birmingham.

[6]   Garcia, J. (2011) ExtJS in Action. Manning Publications, Stamford.

[7]   OpenGeo (2012) Introduction to the OpenGeo Suite. http://presentations.opengeo.org/2012_FOSSGIS/suiteintro.pdf

[8]   Kolios, N. (2009) Spatial Database Postgre SQL/Post GIS and GIS Quantum GIS. User Guide, Greek Free/Open
      Source Software Society (GFOSS).

[9]   Woychowsky, E. (2006) Ajax: Creating Web Pages with Asynchronous JavaScript and XML. Prentice Hall, Upper
      Saddle River, 432 p.

[10] Marini, J. (2002) The Document Object Model: Processing Structured Documents. McGraw-Hill, New York.

[11] Apache Software Foundation (2010) Apache HTTP Server Version 2.2 Documentation.
      https://httpd.apache.org/docs/2.2/

[12] de la Beaujardiere, J., Ed. (2006) OpenGIS® Web Map Server Implementation Specification. Open Geospatial Consor-
      tium Inc., Wayland.

[13] Open Source Geospatial Foundation (2014) Geoserver User Manual. http://docs.geoserver.org/stable/en/user/

[14] Schut, P., Ed. (2007) OpenGIS® Web Processing Service. Open Geospatial Consortium Inc., Wayland.

[15] Portele, C., Ed. (2007) OpenGIS® Geography Markup Language (GML) Encoding Standard. Open Geospatial Consor-
      tium Inc., Wayland.

[16] Wilson, T. and Burggraf, D., Eds. (2009) OGC® KML Standard Development Best Practices. Open Geospatial Consor-
      tium Inc., Wayland.

[17] Butler, B., Daly, M., Doyle, A., Gillies, S., Schaub, T. and Schmidt, Ch. (2008) The GeoJSON Format Specification:
      http://geojson.org/geojson-spec.html

8

Combining Geographic Information Systems for Transportation and Mixed Integer Linear Programming in Facility Location-Allocation Problems

Silvia Maria Santana Mapa[1], Renato da Silva Lima[2]

[1]Federal Office for Education, Science and Technology of Minas Gerais, Congonhas, Brazil
[2]Federal University of Itajuba, Industrial Engineering and Management Institute, Itajubá, Brazil
Email: silvia.mapa@ifmg.edu.br, rslima@unifei.edu.br

Abstract

In this study, we aimed to assess the solution quality for location-allocation problems from facilities generated by the software TransCAD®, a Geographic Information System for Transportation (GIS-T). Such facilities were obtained after using two routines together: Facility Location and Transportation Problem, when compared with optimal solutions from exact mathematical models, based on Mixed Integer Linear Programming (MILP), developed externally for the GIS. The models were applied to three simulations: the first one proposes opening factories and customer allocation in the state of São Paulo, Brazil; the second involves a wholesaler and a study of location and allocation of distribution centres for retail customers; and the third one involves the location of day-care centers and allocation of demand (0 - 3 years old children). The results showed that when considering facility capacity, the MILP optimising model presents results up to 37% better than the GIS and proposes different locations to open new facilities.

Keywords

Geographic Information Systems for Transportation, Location-Allocation Problems, Mixed Integer Linear Programming, Transportation, TransCAD®

1. Introduction

Location-allocation problems are generally complex problems by involving many variables and data. As problem complexity increases, location studies require new information technologies, which permit systems to be treated with effective integration [1]. Approximate facility location models have been proposed by using tools to help them in space, especially when a geographically referenced database is available. In this case, Geographic Information Systems (GIS) are vital for data collection and data analysis, since they incorporate a sophisticated graphical interface into a geographically referenced database, thus becoming powerful tools for spatial analysis and planning [2]. Geographic Information Systems for Transportation (GIS-T) is a special class of GIS, being used for transportation planning and operating. Among its many features, it has sections for facility location. Nothing more natural, thus, than using GIS-T tools to study the best locations for public or private facilities— e.g. factories, distribution centres (DCs), schools or day-care centers—and the best distribution of these units to customers trying to reduce or offset the costs of displacement and/or transportation.

A variety of software programs may be used to aid in location problems. Most of them, however, end up working as a "*black box*", *i.e.* their solution methods are not clear, usually leading users to make assumptions and believe in the efficiency of such solutions. From previous experiments performed in the location-allocation modules of TransCAD®, a commercial GIS-T package, it was verified that it solves problems, however indirectly within two steps. The first step includes the *Facility Location* routine, which identifies the best facility locations (proposing entering new units or closing existing ones, or both), and proceeds to allocate between supply and demand, but without considering the full operation for the facilities. To enforce such restrictions to maximum capacity, a second step is necessary in which the solution of the *Facility Location* routine (FL routine) is submitted to the *Transportation Problem* routine (TP routine). Thus, the solution of the FL routine becomes the input of the TP routine, which will reallocate the supply to demand at the discretion of maximum capacity for facilities. However, this second routine no longer admits the opening or closing facilities previously generated by the FL routine, being pre-conditioned to its initial configuration. Such factor can also clearly compromise the quality of the final solution, since the choice of opening and/or closing of new facilities is certainly subject to their respective capabilities. Additionally, we must consider both routines work with heuristic algorithms in finding solutions. Consequently, there is no guarantee that the solution found after using the routines is the optimal solution, and one would only know for sure by resolving the problem using an optimisation algorithm.

This is the starting point for this study, in which we aimed at assessing the solution quality for location-allocation problems from facilities generated by TransCAD®. Such facilities were obtained after using the FL and TP routines together, when compared with optimal solutions from exact mathematical model, based on MILP and developed outside of the GIS. The MILP model can locate and allocate facilities to customers, while simultaneously obeying bounds of maximum capacity for such facilities, unlike the use of GIS combined routines.

For this study research methodology, classified as modelling and simulation, for both models, *i.e.* GIS and MILP, we applied three simulations (s1, s2, s3) within different complexity levels. For s1: factory, allocation opening and customer allocation within 18 major São Paulo municipalities. For s2: study of location on distribution centres from wholesalers and allocation of their retail customers in the Brazilian states of Minas Gerais and São Paulo. For s3: day-care center location and demand allocation (for children from zero to three years old) in the city of São Carlos (state of São Paulo, Brazil). Thus, the results from the MILP and GIS models were compared and assessed. This study is structured as follows: Section 2 presents considerations on the paper's theoretical analysis; Section 3 demonstrates the chosen research methodology; Section 4 exhibits the three simulations based on the research methodology of Section 3; Section 5 gives a general assessment of the three simulations; and Section 6 offers conclusions for this study along with bibliographic references.

2. Theoretical Analysis

2.1. Facility Location

Facility location problems involve choosing the best location for one or more facilities within a set of possible locations, to provide a high level of customer service, reduce operating costs, or increase profits. Based on [3], a potential optimum solution is sought, which will cut down total facility and transportation cost.

Although location theory began in the 18th century, few applications were actually developed until the mid-1960s, when the issue grew in interest and a wide range of work began to be developed and published [4]. Since then, facility location problems have been widely dealt with in the literature [5], and this research does not intend to exhaust the subject here, but only to present the most relevant topics to the scope of this research. Reference [5] presents a revision of the literature on location models, classifying them according to the principal models (**Table 1**).

Applications for facility location problems occur in both the private and public sectors, focused on being as close as possible to demand, in order to reduce transport costs, increase the coverage area and the level of demand accessibility or totally reduce the facility costs, either by choosing a location because of the financial cost, or the quantity of facilities to be established.

2.2. Geographic Information Systems (GIS)

According to [6], GIS can be defined as an organised collection of hardware, software, skilled personnel, and geographic data, in order to manage a database, making the insertion, storage, handling, removal, update, assessment, and data visualization, both spatial and non-spatial (attribute data), functioning as a valuable tool in planning and management studies. Understanding GIS only as software is to exclude the crucial role it can play in a broad process of decision making. According to [2], since the 1990s there has been a vast and growing interest in GIS in the academic world, software companies, and among professionals, as a consequence of increased processing capacity, reduced microcomputer costs and increased availability of digital cartographic databases. Due to the specific requirements for transport and the adoption of information technology in this area, researchers have increased their efforts to improve the existing GIS approach. Thus, they enable such effort for transportation studies, giving rise to so-called GIS-T.

According to [1], six interrelated factors stand out for the growing use of GIS: 1) there is a range of GIS software available from commercial vendors and universities; 2) increasing capacity of computers to store and retrieve vast amounts of data within reasonable time and cost parameters; 3) more sophisticated and faster printers to produce high resolution and quality outputs; 4) greater availability of spatial data from affordable private companies and government agencies; 5) expanding the use of remote sensing, which requires systems that can deal with a lot of data; and 6) the emergence of Global Positioning System (GPS), which facilitate collecting spatial data at relatively low cost and high accuracy.

2.3. Applying GIS to Facility Location Problems

As cited by [7], "*the combination of GIS and location science is at the forefront of advances in spatial analysis capabilities, offering substantial potential for continued and sustained theoretical and empirical evolution*". According to [8], several studies have been developed in logistics, facility location and vehicle routing by using a GIS, at times combined with other mathematical techniques, applied to both the public and private sectors. In the case of the private sector, the main objective is to minimize logistics costs. For example, for industries, wholesalers, retailers, and distributors to decide the best place to install a unit (or a production plant, a shop, a storage or DC) one should always take into consideration the market positioning of consumers and suppliers, availability of infrastructure, manpower and various other factors affecting production.

Reference [9] presents a new class of facility location-allocation problems to consider the optimal location of two types of installations, static and transportation, to meet a set region with minimum cost. To illustrate this, consider the location of hospitals as a static installation and location of ambulances as transportation facilities.

Table 1. Classification of location problems (Owen *et al.*, 1998).

Problem	Description
p-median model	Locate p facilities at the vertices of a network and allocate the demand to these facilities to reduce the distance travelled. If the facilities are not capacitated and p is fixed, then it is a p-median problem, in which all vertices are assigned to their closest facility. If p is a decision variable and the facilities are capacitated or non-capacitated, it is set as a Capacitated or Non-Capacitated Facility Location problem, respectively. These models are particularly relevant for the design of logistics and load allocation.
Cover sets	Cover sets are based on distance or maximum acceptable trip time, seeking to lessen the quantity of facilities to ensure some level of client coverage. It assumes a finite set of locations and is typically associated with a fixed budget. It is widely used to locate public services, e.g. health centres, post offices, libraries or schools.
Model centres	Model centres is a mini-max problem whose objective is to reduce the maximum distance between demand points and the nearest facility. Problem solving aims to cover all the demand trying to locate a quantity of facilities, provided that it reduces the distance covered. When the location of the facility is restricted to the network node, this is called a centre of vertex problem. If it is possible to locate the problem anywhere in the network, it is an absolute centre problem. This model is primarily applied to emergency services, e.g. fire stations and ambulance stations.

The location of the transportation facilities is dependent on location of static facilities and demand objects. The authors proposed a new GIS platform combined with a heuristic algorithm, customised to solve the problem. Further, the results are available in a user-friendly graphical interface, using thus many resources in developing and applying GIS in location models. Reference [10] developed a model based on GIS for the location of wind, solar, and hydropower energy systems, related to the temporal characteristics of supply and demand worldwide, focusing on locating sources of energy to meet 21st century global demand.

Other examples of GIS application in studies of location can be seen in the literature, such as: Reference [11] describes the development of a planning procedure based on GIS to assist decision makers in site selection when various accessibility criteria are considered. Reference [12], who integrated ArcGIS® with an expert system to assist in the allocation of bank branches; Reference [13] developed a heuristic method integrated with GIS location-allocation problems for maximum coverage and location of the p-median; Reference [14] applied multi-criteria decision analysis in locating power plants using GIS-T; Reference [15] performed a study on current location and a proposed relocation of state schools in Brazilian municipalities; Reference [2] studied the location of day-care centers and health clinics in urban areas using GIS location models; Reference [16] conducted a study to allocate sewage treatment plants and landfills; Reference [17] located cross-docking centres for a pharmaceutical products network; Reference [18] studied how to allocate consolidation terminals for the less-than-truckload trucking industry in Brazil; Reference [19] developed a manufacturing plants or deposits (or both) global-scale model for transnational corporations.

According to [8], GIS capacity rises considerably when it is combined with the use of Operational Research techniques. Reference [20] suggest that the use of exact programming models that use the branch-and-bound algorithm are limited to finding a global solution, effectively optimal to problems with up to 150 nodes on the network. A branch-and-bound algorithm is a tree search method with intelligent enumeration for candidate solutions to the optimal solution of a problem by performing successive partitions of the space of solutions and cutting the search tree by considering limits calculated along the enumeration. The calculation of bounds for the value of the optimal solution is a key part of a branch-and-bound algorithm, since it is used to limit the growth of the tree. Using more robust techniques such as Lagrangian relaxation with optimisation subgradient, increases the ability to handle more complex optimisation problems, roughly from 800 to 900 nodes on the network. Therefore, exact algorithms are only used to solve problems with a restricted number of variables and constraints.

3. Research Methodology

This mathematical model is classified as static and deterministic, resolved by a variant from the p-median problem, focusing on finding the location of p-facilities, since the total distance between supply and demand centres should be minimised while simultaneously allocating the flows between facilities and customers. An interaction between the GIS and external mathematical model, based on MILP, is proposed which will occur during data input and output. According to [20] and [21], the MILP is the most widely used mathematical method for solving location problems formulated by the p-median method, leading to the optimal solution of the problem.

This research sought to assess the solution quality generated by GIS software, in the combined FL and TP routines, when compared to the optimal solution from the MILP model. Even if the GIS mathematical solution quality is bad in a pessimistic assumption, its great help in data collection and solution finding and presentation phases cannot be denied , such as by using its tools for generating graphics and thematic maps etc. According to [1], exporting data from GIS is not difficult, and will serve as input to another location model, external to GIS, which will solve the problem proposed, and then import the results back to GIS to create the graphical results in map form. The MILP model was implemented by using the optimisation software LINGO®, version 7.0, with interface in an MS Excel® spreadsheet, as detailed in the following equations:

Function objective:

$$\min fo = \sum_i \sum_j C_{i,j} * X_{i,j} * d_j \tag{1}$$

Subject to:

$$\sum_i z_i = p \tag{2}$$

$$X_{i,j} \leq z_i \qquad \forall i, j \tag{3}$$

$$\sum_i X_{i,j} = 1 \qquad \forall j \tag{4}$$

$$\sum_j d_j * X_{i,j} \leq m_i * z_i \qquad \forall i \tag{5}$$

$$z_i \in (0,1) \qquad \forall i \tag{6}$$

$$X_{i,j} \in Z^+ \qquad \forall i, j \tag{7}$$

where:

$X_{i,j}$: is the matrix solution. It indicates the percentage of demand j satisfied by i;

$C_{i,j}$: is the cost matrix, represented by the minimum distances between points i and j;

z_i : is a binary vector, indicating which facilities are opened. If $z_i = 1$, the facility i is opened. Otherwise, if $z_i = 0$, then the facility i is not in operation;

d_j : vector that stores customers' demands j;

m_i : input parameter that defines the maximum capacity of facility i;

p : input parameter that specifies the amount of facilities to be opened.

Equation (1) represents the objective function of the model to decrease shipping costs, depending on the distances between points i (set of facilities) and j (set of customers), weighted by the percentage of demands, as well as the objective function in the GIS, for comparison of results generated between the two models. With respect to restrictions, it is worthy to note that Equation (2) is the constraint that indicates the number of facilities to be opened, equivalent to the number p; (3) represents the restriction for opening facilities, *i.e.* if i is not in operation $(z_i = 0)$, the allocation of demand to the facility must be zero; (4) is the restriction of demand coverage, in which all customers should have the total of their demands met; (5) is the restriction of upper bound for capacity of facilities to be opened; (6) makes the output vector z_i a binary vector; and (7) makes the solution matrix $X_{i,j}$ an integer matrix.

The research methodology used in the research can be classified as scientific research being applied, following a quantitative approach with exploratory objectives and using modelling and simulation as a technical procedure [22]. It was implemented in three simulations, according to the GIS and MILP model, involving different complexity levels. In general, the location-allocation problem will approach how to decrease transportation costs in a logistics network, in which the supply (or demand) centres shall fully meet the demands of customers, whether or not subject to the restriction of the upper bound for capacity of facilities. To support the calculations of the models, a computer with an AMD Sempron 2600, 1.83 GHz and 512 MB of RAM was used, operating under the operating system Microsoft Windows XP Professional.

4. Simulation

4.1. Simulation I—Locating Factories

The problem proposed for this simulation is to perform the location and allocation between supply points (plants) and demand (customers) in São Paulo, including 18 of its major cities. To find the points of supply, the choice will be given in a discrete space, which will include all 18 municipalities as candidates for opening facilities for both models, GIS and MILP.

Starting with the GIS model, the FL routine was initially programmed to locate a single plant, in such a way that the plant would supply the total demand of customers (constituting scenario 1 for simulation). New simulation scenarios are proposed as a new unit is opened to reach the total number of facilities, since the average transport cost is less than 100 Km, featuring a coverage problem, *i.e.* situation from scenario 4 for simulation. Importantly, in the scenarios involving the capacity constraint (in this and all other simulations) the supply centre capacity was considered to be the total demand of customers divided equally among the numbers of supply centres. The scenarios were defined as follows: Scenario 1: opening one facility to meet total demand; Scenario 2: opening one new unit, apart from the existing factory located in scenario 1; Scenario 3: opening two new units, apart from the existing factory located in scenario 1; Scenario 4: opening three new units, apart from the existing factory located in scenario 1.

The *objective function* enabled in all scenarios was to decrease the average transport service cost. The numerical results of the scenarios generated by the GIS model for the FL and TP routines, can be found in **Table 2(a)**. From the assessment of **Table 2(a)**, one can realise that the values of transport service Total Cost decreases as the number of factories increase, so that as customers get closer and in turn reducing transport costs. It is hugely important to assess solutions generated by the TP routine, which relocated demands according to the maximum capacity of factories, enabling fragmentation (customers/cities may be served by more than one factory). Thus, it is easy to notice an increase in the value of Total Cost (in scenarios 2, 3, 4), in which all scenarios are compared with respective solution generated by FL routine. This increase is a consequence of the maximum capacity limits imposed on the facilities offered by the TP routine that, once reached, can no longer allocate demand to that installation. The computer processing time was less than one second in all situations.

To draw a comparison between the GIS and MILP models, the same scenario used in the GIS model was simulated in the MILP model. The MILP model can also solve the problem without considering plant capacity, which was called non-capacitated (NC) routine, running exactly as the FL routine of GIS. When you consider these capabilities, the model was called a capacitated (C) routine, making the location and allocation of facilities to their clients and checking the limits of a maximum capacity of these facilities simultaneously, unlike the GIS. Thus, for comparison, the C/MILP routine corresponds to TP/GIS routine, stressing that the TP output of GIS actually means the combined use of the FL and TP routines. The quantitative results of the MILP model are presented in **Table 2(b)**.

Figure 1 presents the spatial results for scenarios 3 and 4, where plant location solutions proposed by NC and C routines were different. Comparing the solutions generated by the GIS and MILP, it can be seen that at first the two models behave similarly. The solutions produced by FL/GIS were equal to those generated by the NC/MILP in all scenarios. This can be seen by comparing the values of the total cost, available in **Table 2(c)**, for the FL/GIS routine and NC/MILP routine.

4.2. Simulation II—Locating Wholesale Distribution Centres

The objective here is to locate DCs for a wholesale company. The company has only one DC located in the city Poços de Caldas, in southern Minas Gerais, Brazil. The wholesaler has a total of roughly 12,000 customers, 30% of them located in southern Minas Gerais and 70% distributed in the São Paulo countryside, concentrated in Vale do Paraíba and the regions of Campinas and Ribeirão Preto. To simplify the problem, all customers within the same municipality were grouped and had their demands combined into a single point, so the demand points were municipalities connected by motorways. There has been, thus, a total of 311 municipality-customers, each with its own geographical location and total demand. The quantity of municipality candidates to have a DC was restricted to 150 for the GIS and MILP models, chosen among those municipality-customers with the highest demand values (89% of total demand). It was the first significant restriction of the MILP model. Despite the software Lingo not imposing limits on the number of problem variables and constraints (only the processing time that rises exponentially), the Excel spreadsheet interfaced with Lingo has a maximum number of columns (256) to store the cost matrix (distances) among the municipality-customers and candidates to DC location.

The criterion to generate the scenarios was the same as in simulation I. **Table 3(a)** shows its numerical results. The minimum total cost of transport with the current DC (scenario 1) would be 820×10^6 Km. When opening a new DC in the TP/GIS model (scenario 2), the wholesaler may reduce their transportation costs by 38.3% when compared to scenario 1. **Table 3(b)** presents the MILP model results for the same scenarios. In addition to the Total Cost variable, **Table 3(b)** also shows the computational processing time, which rises with the number of DCs to be localised. In all scenarios that included the opening of new DCs (2, 3, 4), the C/MILP routine and NC/MILP routine indicated different cities to open such new DCs. The comparison for GIS and MILP model is shown in **Table 3(c)**.

Again, it is perceived that there was no change in the solutions obtained via the FL/GIS and NC/MILP routines. This is already a clear evidence of the quality of the GIS heuristic solution in its non-capacitated model. For the capacitated models, the C/MILP routine results were 22.6% and 13.8% lower than those obtained by the TP/GIS models for scenario 3 and 4, respectively. **Figure 2** illustrates the different locations obtained by the GIS and MILP models for scenario 3. With the evidence that the difference among the capacitated models' solutions increase according to the problem size, it was opted to simulate a problem of even greater size to confirm this hypothesis, which will herein be presented.

Table 2. Results for simulation I.

(a)

GIS	Features		Facility location (FL)		Transportation problem (TP)	
	Existing factories	New factories	Average cost (Km)	Total cost (Km)	Maximum capacity	Total cost (Km)
1	0	1	209.3	**34,954**	167	**34,954**
2	1	1	142.1	**23,740**	84	**24,034**
3	1	2	117.4	**19,609**	56	**22,177**
4	1	3	93.3	**15,591**	42	**16,913**

(b)

MILP	Features		Non-capacitated (NC)		Capacitated (C)	
	Existing factories	New factories	Computational time	Total cost (Km)	Computational time	Total cost (Km)
1	0	1	<1 s	**34,954**	<1 s	34,954
2	1	1	<1 s	**23,740**	<1 s	24,034
3	1	2	<1 s	**19,609**	<1 s	20,597
4	1	3	<1 s	**15,591**	<1 s	16,660

(c)

Comparison	GIS model		MILP model		Range (%)	
	FL	TP	NC	C	FL *vs* NC	TP *vs* C
	Total cost	Total cost	Total cost	Total cost	Total cost	Total cost
	(Km)	(Km)	(Km)	(Km)	(Km)	(Km)
1	34,954	34,954	34,954	34,954	**0.0**	**0.0**
2	23,740	24,034	23,740	24,034	**0.0**	**0.0**
3	19,609	22,177	19,609	20,597	**0.0**	**7.7**
4	15,591	16,913	15,591	16,660	**0.0**	**1.5**

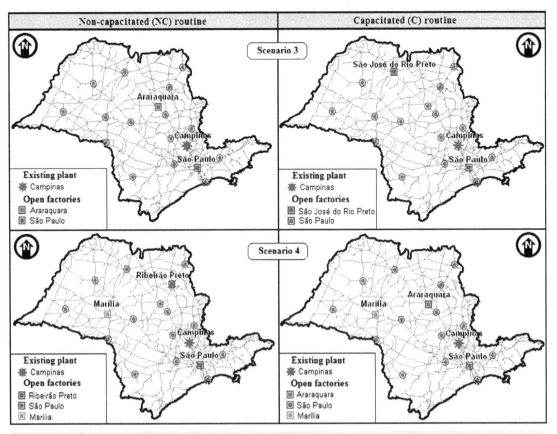

Figure 1. Solutions generated by the MILP model for scenarios 3 and 4.

Table 3. Results for simulation II.

(a)

	Scenarios	GIS model				
		Facility location (FL)			Transportation problem (TP)	
		Open DCs	Average cost (Km)	Total cost (10^6 Km)	Capabilities (Kg)	Total cost (10^6 Km)
1	Allocating the municipality-customers to the existing DC in Poços de Caldas.	A	212	820	3,861,936	820
2	Opening a new DC, before the existing DC in Poços de Caldas.	A, B	121	461	1,930,968	506
3	Opening two new DCs, before the existing DC in Poços de Caldas.	A, B, C	101	383	1,287,312	503
4	Opening three new DCs, before the existing DC in Poços de Caldas.	A, C, D, E	83	315	965,484	402

(b)

	Scenarios	MILP model					
		Non-capacitated (NC)			Capacitated (C)		
		Open DCs	Comp. time	Total cost (10^6 Km)	Open DCs	Comp. time	Total cost (10^6 Km)
1	Allocating the municipality-customers to the existing DC in Poços de Caldas.	A	00:00:01	820	A	00:00:01	820
2	Opening a new DC, before the existing DC in Poços de Caldas.	A, B	00:00:58	461	A, D	00:01:15	500
3	Opening two new DCs, before the existing DC in Poços de Caldas.	A, B, C	00:15:38	383	A, F, G	00:33:57	410
4	Opening three new DCs, before the existing DC in Poços de Caldas.	A, C, D, E	01:44:10	315	A, B, E, F	01:06:05	353

(c)

		GIS model		MILP model		Range (%)	
		FL	TP	NC	C	FL *vs* NC	TP *vs* C
1		820	820	820	820	0.0	0.0
2	Comparison	461	506	461	500	0.0	1.4
3	(Total cost, 10^6 Km)	383	503	383	410	0.0	22.6
4		315	402	315	353	0.0	13.8

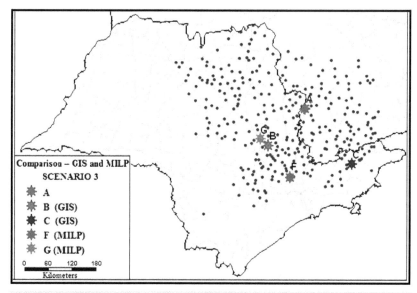

Figure 2. Different solutions of the GIS and MILP model for scenario 3.

4.3. Simulation III—Locating Day-Care Centers

This simulation was based on locating public day-care centers and its demand allocation, *i.e.* children aged from zero to three years old, located in São Carlos city, in the state of São Paulo, Brazil. Data from the Municipal Education Office were used in a geo-referenced GIS file with the city's transport network. These data correspond mainly to address and school attended by all students from public day-care centers (a total of 1014 students), distributed in 10 existing day-care centers (data from 2000).

The first step is to assess the real location and allocation of students and day-care centers. To do so, GIS thematic maps were used, in which students who attend the same day-care center are identified by a same colour. The bad distribution is easily identified by a visual assessment of **Figure 3**, in which the day-care centers appear identified by a single symbol, using the letters from A to J to distinguish them. In the map's legend, every day-care center is associated with the number of students enrolled. From information obtained in the real situation assessment, different location scenarios and allocation among day-care centers and students have been proposed, seeking to concentrate more students living by day-care center they attend. The scenarios was named as *Real* (real locations and allocations between students and day-care centers), *Scenario* 1 (reallocation of students for the ten-existing day-care centers), *Scenario* 2 (ten-existing day-care centers, opening a new unit), *Scenario* 3 (ten-existing day-care centers, opening two new units) and *Scenario* 4 (closing one of the ten day-care centers and opening a new unit).

For scenario 1, the capacity limits imposed on each of the day-care centers for the TP routine were established as the number of students actually enrolled in a specific day-care center at the time of collection and data analysis. By streamlining, the lack or excess of vacancies was not considered, since the aim is only to compare the GIS and MILP model. Thus, the same parameters in both models were maintained. For scenarios 2 and 3, in which we proposed to open new day-care centers, the capacity of the new units was established as 10% from the total demand of students (101 from 1014 students enrolled), lessening the capacity for existing day-care centers. In scenario 4, the supply capacity for closed day-care centers was reallocated to the new unit opened by the model. Relatively few candidates for locations were chosen by using the location of equidistant points (0.7 Km), covering the entire city for a total of 120 points. Added to the ten-existing day-care centers, there are 130 candidates, a set used for both models (GIS and MILP). Once again, the processing capacity of the MILP exact model was limiting for the number of candidates considered.

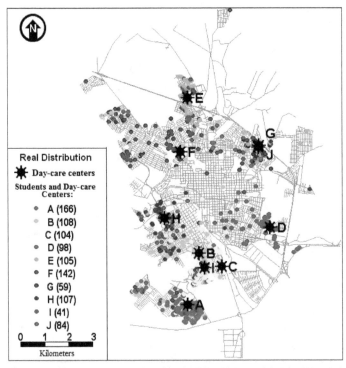

Figure 3. Real distribution of day-care centers and their students.

The GIS model results are shown in **Table 4**. Based on mean, maximum and total cost, one can see that all the simulated scenarios showed better results than the Real situation, with reductions of up to 60%. A simple reallocation of students in the existing model (scenario 1) would lead to reductions of 50% in transportation costs.

The MILP model results are presented in **Table 5**. The results are quite similar to the GIS model in quantitative terms, with significant reductions in displacement costs for the real situation. The most significant finding, however, is that in scenarios 2, 3, 4 the NC routine and C routine of the MILP model generated different locations within the same scenario, *i.e.* when the restriction to maximum capacity is imposed on the model, location solutions for day-care centers, and consequently the allocation of students, are different. For example, in scenario 2 of the NC model, the day-care center to be opened would be letter K; and for C model would be letter M. The highest computational processing was nearly 2 hours in scenario 3 capacitated (01:52:14).

Table 4. Results for the GIS model.

		Real Situation	Scenario 1 FL	var %	TP	var %	Scenario 2 FL	var %	TP	var %
DAY-CARE CENTER	A	166	196	18	166	0	196	18	149	−10
	B	108	115	6	108	0	115	6	97	−10
	C	104	22	−79	104	0	22	−79	94	−10
	D	98	78	−20	98	0	78	−20	88	−10
	E	105	105	0	105	0	105	0	95	−10
	F	142	147	4	142	0	69	−51	128	−10
	G	59	37	−37	59	0	37	−37	53	−10
	H	107	88	−18	107	0	88	−18	96	−10
	I	41	125	205	41	0	125	205	37	−10
	J	84	101	20	84	0	101	20	76	−10
	K						78		101	
	L									
Average cost (Km)		2.04	0.99	−52	1.10	−46	0.86	−58	1.23	−40
Maximum cost (Km)		10.52	5.28	−50	5.34	−49	5.20	−51	5.20	−51
Total cost (Km)		2070	1001	−52	1120	−46	877	−58	1244	−40

		Real Situation	Scenario 3 FL	var %	TP	var %	Scenario 4 FL	var %	TP	var %
DAY-CARE CENTER	A	166	169	2	133	−20	196	18	166	0
	B	108	115	6	86	−20	119	10	108	0
	C	104	22	−79	83	−20	Closed	-	Closed	-
	D	98	78	−20	78	−20	78	−20	98	0
	E	105	105	0	84	−20	105	0	105	0
	F	142	69	−51	114	−20	69	−51	142	0
	G	59	37	−37	47	−20	37	−37	59	0
	H	107	88	−18	86	−20	88	−18	107	0
	I	41	125	205	34	−17	143	249	41	0
	J	84	101	20	67	−20	101	20	84	0
	K		78		101		78		104	
	L		27		101					
Average cost (Km)		2.04	0.81	−60	1.09	−47	0.87	−57	1.43	−30
Maximum cost (Km)		10.52	5.20	−51	5.20	−51	5.20	−51	5.57	−47
Total cost (Km)		2070	822	−60	1102	−47	885	−57	1452	−30

Table 5. Results for the MILP model.

		Real	Scenario 1				Scenario 2			
		Situation	NC	var %	C	var %	NC	var %	C	var %
	A	166	196	18	166	0	196	18	149	−10
	B	108	115	6	108	0	115	6	97	−10
	C	104	22	−79	104	0	22	−79	94	−10
	D	98	78	−20	98	0	78	−20	88	−10
DAY-CARE CENTER	E	105	105	0	105	0	105	0	95	−10
	F	142	147	4	142	0	69	−51	128	−10
	G	59	37	−37	59	0	37	−37	53	−10
	H	107	88	−18	107	0	88	−18	96	−10
	I	41	125	205	41	0	125	205	37	−10
	J	84	101	20	84	0	101	20	76	−10
	K						78			
	L									
	M								101	
	N									
Average cost (Km)		2.04	0.99	−52	1.10	−46	0.86	−58	1.08	−47
Maximum cost (Km)		10.52	5.28	−50	5.34	−49	5.20	−51	5.28	−50
Total cost (Km)		2070	1001	−52	1120	−46	877	−58	1095	−47

		Real	Scenario 3				Scenario 4			
		Situation	NC	var %	C	var %	NC	var %	C	var %
	A	166	169	2	133	−20	196	18	166	0
	B	108	115	6	86	−20	119	10	108	0
	C	104	22	−79	83	−20	closed	-	closed	-
	D	98	78	−20	78	−20	78	−20	98	0
DAY-CARE CENTER	E	105	105	0	84	−20	105	0	105	0
	F	142	69	−51	114	−20	69	−51	142	0
	G	59	37	−37	47	−20	37	−37	59	0
	H	107	88	−18	86	−20	88	−18	107	0
	I	41	125	205	34	−17	143	249	41	0
	J	84	101	20	67	−20	101	20	84	0
	K		78		101		78			
	L		27							
	M								104	
	N				101					
Average cost (Km)		2.04	0.81	−60	1.01	−50	0.87	−57	1.05	−49
Maximum cost (Km)		10.52	5.20	−51	5.20	−51	5.20	−51	5.34	−49
Total cost (Km)		2070	822	−60	1027	−50	885	−57	1063	−49

The comparative assessment between the GIS and MILP models is shown in **Table 6**. In all scenarios, both models came to the same results for non-capacitated routines. Assessing the solutions for the capacitated routines, however, it can be seen that the C/MILP routine reached better solutions than the TP/GIS routine, in scenarios 2, 3, 4, with differences up to 37% (scenario 4). Concerning the chosen locations, the NC/MILP model located the day-care centers in the same places as the FL/GIS model (because of null values in the penultimate column of **Table 6**). By contrast, the C/MILP model generated different locations for the TP/GIS model in all

Table 6. Comparing the GIS and MILP models.

Scenario	Costs (Km)	GIS		MILP		var (%)	
		FL	TP	NC	C	FL *vs* NC	TP *vs* C
1	Average cost	0.99	1.10	0.99	1.10	0	0
	Maximum cost	5.28	5.34	5.28	5.34	0	0
	Total cost	1001	1120	1001	1120	0	0
2	Average cost	0.86	1.23	0.86	1.08	0	14
	Maximum cost	5.20	5.20	5.20	5.28	0	−2
	Total cost	877	1244	877	1095	0	14
3	Average cost	0.81	1.09	0.81	1.01	0	7
	Maximum cost	5.20	5.20	5.20	5.20	0	0
	Total cost	822	1102	822	1027	0	7
4	Average cost	0.87	1.43	0.87	1.05	0	37
	Maximum cost	5.20	5.57	5.20	5.34	0	4
	Total cost	885	1452	885	1063	0	37

scenarios, where opening or closing (or both) of new units was proposed. In scenario 4, for example, the GIS model located the day-care center K to the north; and MILP model located day-care center M to the south with 37% lower cost.

5. General Assessment for the Results

It should be stressed here that a heuristic procedure is associated with the FL/GIS routine, and that the NC/MILP model is an exact model, ensuring the optimal solution for the problem addressed. The heuristic method subjected to the FL/GIS routine produced solutions with the same locations and total cost values the exact solution of NC/MILP routine used. Therefore, the three simulations proposed an important conclusion is for the heuristic performance of the FL routine, classifying it as good, since it converged to the optimal mathematical solution, in all the simulated scenarios.

Besides having been shown to be a robust heuristic, the FL/GIS routine, when performing location and allocation, enhances the capacity of facilities according to the demand location. Quoting an example, for scenario 1 in simulation III, to which it sought to relocate students to existing day-care centers, ignoring the opening or closing units, FL routine generated the best allocation between students and respective day-care centers as a result, as indicated on **Table 4**. Thus, day-care centers should have their capabilities designed (or re-designed) to meet the number of students allocated, as indicated by the FL routine. If the capabilities of all day-care centers were scaled to the values indicated in FL routine, the GIS model would decrease overall costs of transport by roughly 52%, without having to open new units.

For simulations I, II, III, assessing the final solutions for scenarios with opening or closing new units, the MILP model always yielded better results than the GIS when considering facility capacity. The lower transportation costs, through the opening of units in different locations, can be explained by the fact that the GIS model performs location and respective allocation (according to the capacity limits) in an indirect way, demonstrating itself as a disadvantage for the GIS model. As the facility capacity's upper bound restriction is relaxed (or rather, instead of fixing an upper bound, capacity ranges are established, thus permitting an increase in capacity), the solutions generated by the model GIS tend to approach the MILP solutions for suppler-customer allocations. The restriction is relaxed until the problem becomes non-capacitated, in which the GIS solution equals the MILP solution, generating the same locations and allocations and, consequently, the same total cost values for transportation. It occurs due to the many chances that the GIS model will have to relocate the demand to supply, previously located by the FL routine, since facilities will possess an upper bound of greater capacity, allowing the allocation for more demands. **Table 7** illustrates such fact, presenting the simulation II for location between DCs and its customers.

For simulation II, by increasing the capacity range limit by 20% of its initial value and performing the capable

Table 7. Solutions to increase the upper bound of capacity.

Scenario	GIS				MILP			Range (%)
	Capacity limit	Places for facilities	Demand allocation	Total cost	Places for facilities	Demand allocation	Total cost	TP *vs* MILP
1	4,634,323	A	3,861,936	820	A	3,861,936	820	0.0
2	2,317,162	A	1,544,774	478	A	1,544,774	478	0.0
		B	2,317,162		B	2,317,162		
3	1,544,774	A	1,544,774	460	A	941,708	396	16.2
		B	1,544,774		F	1,544,774		
		C	772,388		H	1,375,454		
4	1,158,581	A	1,017,463	366	A	869,718	336	8.9
		C	527,311		F	1,158,581		
		D	1,158,581		B	1,153,346		
		E	1,158,581		E	680,291		

routines in GIS models, new solution allocations are obtained between DCs and customers. **Table 6** shows this, when comparing it with the solution of **Table 3(b)**. Such different solution allocations consequently generate different total cost values for service. Although there are still differences among location and allocation solution models generated by GIS and MILP with increased capacity limits, the variance between the two models decreased. Considering the limit values imposed previously for scenario 2's simulation, for example, the variation between the two models was 1.4% (**Table 3(c)**). By contrast, with a 20% higher capacity limit, there was no difference between the two models. One may also find it for scenarios 3 and 4 of this simulation, and these tests for capacity bounds were also performed for simulation I, III. Thus, it can be said that the findings were the same.

Further, it must be considered that solution for location-allocation problems is strongly related to the set of candidate locations. If this set changes, the solution tends to change. Thus, the greater the number of candidate sites, the better will be the solution's expected quality. However, a rise in the problem's variables is proportional to its complexity, mirroring the rise of computational processing time consumed for the solution generation. Therefore, the GIS model has the advantage of being heuristic, and it allows working with multiple variables in a relatively short computer processing time. For all simulated scenarios, the GIS always generated solutions in less than five seconds. In the MILP model based on an optimisation algorithm, the processing time was much higher, nearly consuming a maximum consumption time of two hours in the routine with capacity constraints. To compensate this computer processing time, which rises exponentially, the solution is to decrease the variables, which in fact had to be done for larger simulations, *i.e.* II, III.

6. Conclusions

In this study, the solution quality for location-allocation problems from facilities generated by TransCAD®, a GIS-T was assessed. Such facilities were obtained after using FL and TP routines together, when compared with optimal solutions from exact mathematical models, based on MILP, developed apart from the GIS. The resulting analysis showed different situations when the restrictions imposed by the capacity of facilities were considered (or not). For the non-capacitated routines, the GIS and MILP models presented exactly the same solutions. It can be concluded that the heuristic method of GIS for the FL routine is quite efficient. When considering the capacitated routines, however, resolving the problem with the GIS combined model (TP routine after FL routine) leads to results different from those of the MILP optimisation routine, indicating the opening of facilities in different locations. As the transportation costs for the MILP model are up to 37% lower, it can be concluded that such performance is better, probably due to the simultaneous resolution of the phases of facility location and demand allocation.

Paradoxically, it is shown that GIS possesses the ability to handle larger problems with more variables and candidates involved, as their routines are based on heuristic methods. The optimisation solution of the MILP model is strongly conditioned to the problem size to be solved, as observed by the processing time. For future work, thus, the authors suggest the development of a new model apart from GIS. The simultaneous solution of phases to locate facilities and allocate demands through the use of heuristic algorithms should also be considered.

Acknowledgements

The authors would like to express their gratitude to the Brazilian agencies CNPq (National Council for Scientific and Technological Development), and FAPEMIG (Foundation for the Promotion of Science of the State of Minas Gerais), which have been supporting the efforts for the development of this work in different ways and periods.

References

[1] Church, R.L. (2002) Geographical Information Systems and Location Science. *Computers & Operations Research*, **29**, 541-562. http://dx.doi.org/10.1016/S0305-0548(99)00104-5.

[2] Lima, R.S., Silva, A.N.R. and Mendes, J.F.G. (2003) A SDSS for Integrated Management of Health and Education Facilities at the Local Level: Challenges and Opportunities in a Developing Country. *Proceedings of 8th International Conference on Computers in Urban Planning and Urban Management*, Center for Northeast Asian Studies, Tohoku University, Sendai.

[3] Ballou, R.H. (2004) Business Logistics/Supply Chain Management. 5th Edition, Pearson Education International, New Jersey, 789 p.

[4] Owen, S.H. and Daskin, M.S. (1998) Strategic Facility Location: A Review. *European Journal of Operational Research*, **111**, 423-447, http://dx.doi.org/10.1016/S0377-2217(98)00186-6 .

[5] Pizzolato, N.D., Barros, A.G., Barcelos, F.B. and Canen, A.G. (2004) Localização de escolas públicas: síntese de algumas linhas de experiências no Brasil. *Pesquisa Operacional*, **24**, 111-131. http://dx.doi.org/10.1590/S0101-74382004000100006.

[6] Lima, R.S., Silva, A.N.R., Egami, C.Y. and Zerbini, L.F. (2000) Promoting a More Efficient Use of Urban Areas in Developing Countries: An Alternative. *Transportation Research Record: Journal of the Transportation Research Board*, **1726**, 8-15. http://dx.doi.org/10.3141/1726-02.

[7] Murray, A. (2010) Advances in Location Modeling: GIS Linkages and Contributions. *Journal of Geographical Systems*, **12**, 335-354. http://dx.doi.org/10.1007/s10109-009-0105-9.

[8] Lorena, L.A.N., Senne, E.L.F., Paiva, J.A.C. and Pereira, M.A. (2001) Integration of Location Models to Geographical Information Systems. *Gestão e Produção*, **8**, 180-195. http://dx.doi.org/10.1590/S0104-530X2001000200006.

[9] Gu, W., Wang, X. and Geng, L. (2009) GIS-FL Solution: A Spatial Analysis Platform for Static and Transportation Facility Location Allocation Problem. In: *Proceedings of 18th International Symposium on Methodologies for Intelligent Systems*, Springer LNAI, Prague, 453-462.

[10] Biberacher, M. (2008) GIS-Based Modeling Approach for Energy Systems. *International Journal of Energy Sector Management*, **2**, 368-384. http://dx.doi.org/10.1108/17506220810892937 .

[11] Tong, D., Lin, W.H., Mack, J. and Mueller, D. (2010) Accessibility-Based Multicriteria Analysis for Facility Siting. *Transportation Research Record: Journal of the Transportation Research Board*, **2174**, 128-137. http://dx.doi.org/10.3141/2174-17

[12] Oliveira, R.L., Lima, R.S and Lima, J.P. (2013) Arc Routing Using a Geographic Information System: Application in Recyclable Materials Selective Collection. *Advanced Materials Research*, **838-841**, 2346-2353. http://dx.doi.org/10.4028/www.scientific.net/AMR.838-841.2346

[13] Arakaki, R.G.I. and Lorena, L.A.N. (2006) A Location-Allocation Heuristic (LAH) for Facility Location Problems. *Production*, **16**, 319-328. http://dx.doi.org/10.1590/S0103-65132006000200011

[14] Zambon, K.L., Carneiro, A.A.F.M., Silva, A.N.R. and Negri, J.C. (2005) Multicriteria Decision Analysis for Site Selection of Thermoelectric Power Plants Using GIS. *Pesquisa Operacional*, **25**, 183-199. http://dx.doi.org/10.1590/S0101-74382005000200002

[15] Pizzolato, N.D. and Silva, H.B.F. (1997) The Location of Public Schools: Evaluation of Practical Experiences. *International Transactions in Operational Research*, **4**, 13-22. http://dx.doi.org/10.1111/j.1475-3995.1997.tb00058.x

[16] Naruo, M.K. (2003) O estudo do consórcio entre os municípios de pequeno porte para disposição final de resíduos sólidos urbanos utilizando Sistemas de Informações Geográficas. Dissertation, University of Sao Paulo (USP), Sao Paulo.

[17] Dobrusky, F.G. (2003) Optimal Location of Cross-Docking Centres for a Distribution Network in Argentina. Dissertation, Massachusetts Institute of Technology (MIT), Cambridge, Massachusetts.

[18] Friend, J.D. and Lima, R.S. (2011) From Field to Port: The Impact of Transportation Policies on the Competitiveness of Brazilian and US Soybeans. *Transportation Research Record: Journal of the Transportation Research Board,* **2238,** 61-67. http://dx.doi.org/ 10.3141/2238-08

[19] Hamad, R. (2006) Modelo para localização de instalações em escala global envolvendo vários elos da cadeia logística. Dissertation, University of Sao Paulo (USP), Sao Paulo.

[20] Church, R.L. and Sorensen, P. (1996) Integrating Normative Location Models into GIS: Problems and Prospects with the *p*-Median Model. In: Longley, P. and Batty, M., Eds., *Spatial Analysis: Modelling in a GIS Environment,* GeoInformation International, Cambridge, 167-183.

[21] Vallim Filho, A.R.A. (2004) Localização de centros de distribuição de carga: contribuições à modelagem matemática. Ph.D. Thesis, University of Sao Paulo (USP), Sao Paulo.

[22] Bertrand, J.W.M. and Fransoo, J.C. (2002) Operations Management Research Methodologies Using Quantitative Modeling. *International Journal of Operations & Production Management,* **22,** 241-264. http://dx.doi.org/10.1108/01443570210414338

9

A Model for Measuring Geographic Information Systems Success

Khalid A. Eldrandaly, Soaad M. Naguib*, Mohammed M. Hassan

Information Systems Department, Faculty of Computers and Informatics, Zagazig University, Al-Sharqiyah, Egypt
Email: Khalid_eldrandaly@zu.edu.eg, *so3ad_mn@zu.edu,eg, monirhm2002@yahoo.com

Abstract

Geographic Information Systems (GIS) have become a fact of our life as they are being used by more people and organizations for more complex decision problems than ever before. The use of GIS can achieve valuable benefits for individuals, organizations and society; however, the achievement of these benefits depends on the success of GIS. While information systems (IS) success models have received much attention among researchers, there is a general scarcity of research conducted to measure the GIS success. This paper proposes a success model for measuring GIS success by extending and modifying previous IS success models. The developed success model consists of two main levels: GIS project diffusion success, and GIS post-implementation success. The first level identifies the critical success factors (CSFs) that influence the success of GIS adoption at each stage of the diffusion process. The second level of the proposed model identifies and organizes the success dimensions (outcome measures) of GIS in temporal and causal relationships. In order to assess the relationships among the success dimensions, 11 hypotheses were tested. Data were collected through a questionnaire that was distributed to 252 GIS users/managers in Egypt and abroad. The empirical results support 6 hypotheses and reject 5 hypotheses.

Keywords

IS Success Models, GIS Success Model, Critical Success Factor, Model Validation

1. Introduction

Geographic Information Systems (GIS) are a mainstream technology with a vital and growing use across all industries [1]. A GIS is a computer-based information system that enables capture, modeling, storage, retrieval, sharing, manipulation, analysis, and presentation of geographically referenced data [2]. A working GIS inte-

---

*Corresponding author.

grates five key components: hardware (the equipment needed to support the many activities of GIS ranging from data collection to data analysis and sharing), software (different GIS software packages for creating, editing and analyzing data), data (the core of any GIS, categorized as spatial and non-spatial data), organizational structure and people (well-trained and skilled people to use and maintain the GIS) and methods (well-designed plan and business rules that are the models and operating practices unique to each organization) [3]. GIS are well established as giving competitive advantage and enhancing organizational decision-making in a wide array of functions, including improved information sharing and flows, better informed decision making, stronger competitive ability, greater analysis and understanding of problems, justification for decision made, improved visualization of data, cost saving, increased effectiveness, and better quality output [4]. However, the achievement of these valuable benefits depends on the success of GIS. Although measuring traditional IS success has been a major topic in IS research, there is a scarce of research addressing GIS success. This paper accordingly attempts to propose a comprehensive multidimensional success model for GIS and to empirically investigate the multidimensional relationships among the success measures. The validated GIS success model can provide GIS managers with a useful framework for evaluating GIS success.

This paper is structured as follows. First, we review the development of IS success models, and consider the challenges and difficulties facing these IS success models. Second, based on prior studies, a GIS success model and a comprehensive set of hypotheses are proposed. Third, the methods, measures and results of the study are presented. And, finally, the results are discussed.

2. IS Success Models

Information system (IS) success is one of the most researched topics in IS literature [5]. Measuring IS success is believed to be a critical issue in the IS field. Several studies have been conducted, and considerable attention was paid to this issue; due to the amounts of money, time and efforts spent on IT/IS projects.

The IS literature provides several definitions and measures of IS success. As [6] stated that, although there are nearly as many measures as there are studies; obviously, there is no ultimate definition of IS success. The definition of IS success may vary in respect of different types of IS that yield different benefits for individuals, workgroups and organizations [7].

2.1. The Delone and Mclean IS Success Model (1992)

A wide range of research has proposed IS success models [6] [8]-[10]. These models postulate their own definitions of IS success and factors that affect the defined IS success, however, a first synthesis of the manifold perspectives on IS success and its underlying antecedents was achieved by DeLone and McLean (1992) [6], who developed an IS success model. This model has received much attention from IS scholars and can be regarded as one of the most prominent and influential models in IS research [11]. DeLone and McLean (1992) [6] made a major breakthrough. They conducted a comprehensive review of IS success literature published between 1981-1987 in seven journals in the IS field, and proposed a model of IS success shown in **Figure 1**.

2.2. The 3-D Model of IS Success

Since 1992, a number of studies have undertaken empirical investigations of the multidimensional relationships among the measures of IS success such as [12] [13]. In addition to studies that have tested and validated the

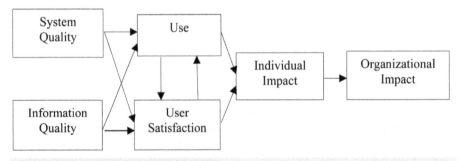

Figure 1. DeLone and McLean model (1992).

DeLone and McLean IS success model (1992), several studies have been published that challenge, critique, or extend the model itself. On balance, these articles have contributed to a better understanding of IS success such as Ballantine *et al.* (1996) [18].

Ballantine *et al.* (1996) [18] assessed the DeLone and McLean model (1992) from a number of different viewpoints and a new model, the 3-D model, has been proposed. First, the independent variable or the critical success factors, factors that may cause success rather than being part of success, have been incorporated to the 3-D model. Second, the causal relationship that exist between the success dimensions in DeLone and McLean model (1992) has been criticized arguing that, success in one dimension does not lead directly to success at the next dimension. If this were so, then one need only be successful at system quality to ensure the success of the whole IS. Finally, Ballantine *et al.* (1996) [18] mentioned that DeLone and McLean (1992) [6] provide a unidirectional model which moves from system and information quality to organizational impact of IS, the impact upon organizational learning is not discussed. The implication in the DeLone and McLean model (1992) is that success is necessary at each stage otherwise the next stage will not be successful. It is not sufficient to move only in one direction. Unless individuals and the organization can learn from experience and develop better systems and recognize better information quality then it is unlikely that the measurement of IS outcome will serve any useful purpose. Thus, Ballantine *et al.* (1996) [8] depict the importance of organization learning through time and experience in their model through the learning feedback loop as shown in **Figure 2**. The 3-D model separates success into three fundamental levels: the technical development level, the deployment to the user, and the delivery of business benefits. Hence it is termed the 3-D model as shown in **Figure 2**. In the 3-D model, filters act between the levels of IS success and contains influences which inhabit or encourage the adoption of the IS at the next higher level. Success at each level is influenced by a number of different independent variables and the outcomes of each level (development, deployment, and delivery) are the closest equivalent to dependent variable (outcome measures of IS success).

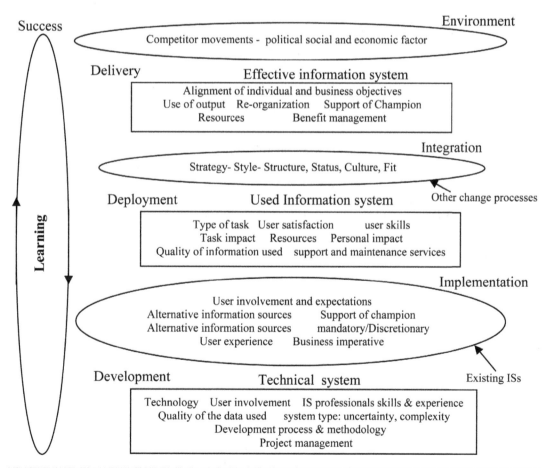

Figure 2. The 3-D model of IS success.

2.3. The Update DeLone and McLean Model (2003)

Based on suggestions offered by researchers, and criticisms directed to the DeLone and McLean original model, DeLone and McLean (2003) [9] proposed an updated version of their IS success model (1992) and evaluated its usefulness in light of the dramatic changes in IS practice, especially the advent and explosive growth of e-commerce. The updated model is shown in **Figure 3**. The primary differences between the original and updated models include:

1) The addition of service quality to reflect the importance of service and support in successful e-commerce systems;
2) The addition of intention to use to measure user attitude;
3) The collapsing of individual impact and organizational impact into the net benefits construct.

The categories of the updated taxonomy were system, information, and service quality, intention to use, use, user satisfaction, and net benefits. DeLone and McLean models (1992, 2003) could serve as a basis for the selection of appropriate IS measures. Researchers had to choose several appropriate success measures based on the objectives and the phenomena under investigation, as well as consider possible relationships among the success dimensions when constructing the research model [14].

2.4. Challenges and Difficulties Facing the IS Success Models

While the updated DeLone and McLean model (2003) is a comprehensive IS success model, it suffers from certain difficulties. First, the Net Benefit measure in the model is conceptually too broad to define. As [15] suggests, "The new net benefits construct immediately raises three issues that must be addressed: what qualifies as a benefit? for whom? and at what level of analysis" (p. 32). Thus, when using the updated DeLone and McLean model, researchers need to clearly and carefully define the stakeholders and the context in which Net Benefits are to be measured.

Second, as DeLone and McLean (1992) [6] stated "The selection of IS success measures should also consider the contingency variable such as the independent variable being researched: the organizational strategy, structure, size, and environment of the organization being studied; the technology being employed; and the task and individual characteristics of the system under investigation" (p.88). Thus, DeLone and McLean (1992) [6] recognized the limited perspective of their model. They identify only the dependent variables (outcome measures) of IS success.

While the 3-D IS success model represents a holistic view of the concept of IS success. This model did not focus on the outcome measures of IS as DeLone and McLean (1992, 2003) did. Although Ballantine *et al.* (1996) [8] incorporated the critical success factors (independent variables) to their success model, the outcome measures of IS success are unclear and misspecified.

Based on the above mentioned literature, this study proposes a new GIS success model by extending and respecifying both the 3-D model of IS success and the updated DeLone and McLean model (2003) in the context of GIS.

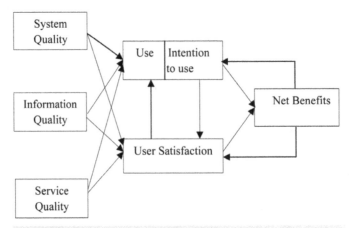

Figure 3. DeLone and McLean model (2003).

3. Research Model and Hypothesis

The literature search indicated that, there is a general scarcity of models and frameworks for measuring GIS success. However, there are some frameworks that were developed for evaluating the contributions of GIS to efficiency, effectiveness, and societal well being (see e.g., [16]-[19]).

By extending and respecifying both the 3-D model and the updated DeLone and McLean model (2003) of IS success, we proposes a GIS success model that provides a holistic view of GIS success concept via defining GIS success as a cumulative process that starts from initiating successful GIS projects and ending with the success of GIS in delivering their business objectives. Thus, we divide success into two levels: (1) GIS diffusion success, and (2) GIS post-implementation success, as shown in **Figure 4**.

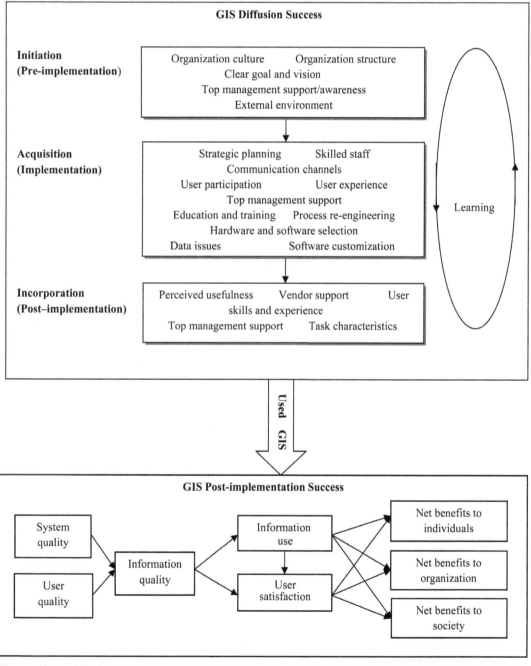

Figure 4. The proposed GIS success model.

3.1. Level One: GIS Diffusion Success

This level extends and respecifies the 3-D model of IS success developed by Ballantine *et al.* (1996) [8] to GIS context by identifying the CSFs in each stage of GIS diffusion process (instead of development, deployment, and delivery levels in the 3-D model) that influences the success of GIS project adoption at each stage. The concept of CSFs was popularized in the context of IS and project management (PM) by [20] as factors affecting the success of activities and projects. CSF is defined by [21] as "the limited number of areas in which satisfactory results will ensure successful competitive performance for the individual, department, or organization. CSFs are the few key areas where 'things must go right' for the business to flourish and for the managers goal to be attained" (p. 385). CSFs make it easier for managers to prioritize vital aspects of a project [22].

According to [23] diffusion is "the fundamental process that is responsible for the transfer of innovations from the workshops of their inventors to becoming a daily part of the lives of a large section of society". GIS project diffusion can be classified into three main stages [24]-[27]: (1) initiation (pre-implementation), (2) acquisition (implementation), and (3) incorporation (post-implementation). Detailed descriptions of these stages and the steps carried out in each one are reported elsewhere (see, e.g., [28] [29], for an excellent overview). The main objective of the first level of the proposed GIS success model is to identify the CSFs in each stage of the GIS project diffusion that influence the success of GIS project adoption at each stage as shown in **Table 1**. Some CSFs may work at more than one stage. The CSFs identified in the three stages of GIS diffusion process were extracted from:

1) GIS success research cited in the literature which is mostly based on case studies or observations of GIS projects and practices, such as [30]-[32];
2) GIS failure research which is based on lessons learned from certain types of GIS projects, but they are mostly similar enough to be generalized, such as [33]-[35];
3) Researches about GIS implementation that mentioned CSFs briefly such as [36]-[38].

Table 1. GIS CSFs according to their occurrence in GIS diffusion stages.

Stage	GIS CSFs	Sources
Initiation (Pre-Implementation)	Organization Culture	[23]-[26] [37] [38] [40]-[42]
	Organization Structure	[30] [37] [38] [40] [43]-[45]
	Clear Goal and Vision	[24] [30] [33] [37] [45]-[48]
	Top Management Support/Awareness	[19] [30]-[33] [37] [38] [40] [45]-[55]
	External Environment	[40] [44]-[46] [56]
Acquisition (Implementation)	Strategic Planning	[33] [34] [37] [40] [42] [46] [48] [49] [53] [55] [57]
	Skilled Staff	[30] [33] [35] [38] [40] [42] [43] [45] [46] [51] [52] [55]
	Communication Channels	[19] [26] [30] [34] [37] [38] [40] [52] [53] [56]
	User Participation	[19] [30] [32] [33] [36]-[38] [40] [43] [46] [49] [52] [53]
	Education and Training	[19] [30]-[34] [36]-[38] [40] [43]-[49] [52]-[55] [58] [59]
	Business Process Re-Engineering	[30] [33] [49] [58]
	Hardware and Software Selection	[19] [33] [35] [38] [40] [47] [49] [52]
	Software Customization	[34] [35]
	Data Issues	[19] [30] [32]-[34] [42] [45]-[47] [51] [52] [54] [58] [60]
Incorporation (Post-Implementation)	Perceived Usefulness	[19] [26] [38] [45] [47] [56]
	Vendor Support	[19] [26] [32] [35] [37] [45] [53]
	User Skills and Experience	[19] [30] [31] [35] [36] [38] [44] [47] [49] [52] [56]
	Task Characteristics	[60] [61]

3.1.1. The Initiation Stage (GIS Concept Introduction and Funding Commitment)

The objective of this stage is to learn about GIS technology, explore the appropriateness of GIS for the organization, and to gain official sanction for the next stage (acquisition). The result from this stage reveals if the organization is ready to accept GIS. This stage includes the following CSFs: organization culture, organization structure, clear goal and vision, top management support/awareness, and the external environment [24].

3.1.2. The Acquisition Stage (GIS Needs Analysis, Design, and Installation)

The acquisition stage begins when the organization becomes aware of the GIS and decides to adopt it, and when GIS advocates have confidence that financial and management support exist to establish a budget. In this stage, the organization engages in the activities necessary to put the GIS into practice [26] [29]. Planning, system design, development, installation and operating the system are the objectives of this stage. This stage includes the following CSFs: strategic planning, skilled staff, communication channels, user participation, user experience, education and training, process re-engineering, hardware and software selection, data issues, software customization, and top management support.

3.1.3. The Incorporation Stage (GIS Acceptance and Use)

The incorporation stage focuses on the acceptance of the technology by members of organization and its utilization over time [39]. All operational GIS installations must move into this third stage if the promises of GIS technology are to be realized. The objective of this stage is to use GIS technology for the daily tasks and decision making required by the organization [24]. This stage includes the following CSFs: perceived usefulness, vendor support, user skills and computer experience, task characteristics and top management support.

The stages of GIS diffusion process are represented in the proposed success model with respect to their occurrence. Hence, the arrows between stages of GIS diffusion process represent temporal relationship.

During the achievement of GIS diffusion success, as shown in **Figure 4**, there is an issue of learning through time and experience which is suggested by Ballantine *et al.* (1996) [8] in his IS success model. Thus, many factors which have an effect on the success of the three stages of GIS diffusion, will be probably affected by previous experience and the maturity of the organization in developing IS projects. Furthermore, if the organization is a learning organization, it will have in place procedures and people who monitor the success and adapt changes in the diffusion processes in order to achieve success. This is represented by the learning feedback loop in **Figure 4**.

3.2. Level Two: GIS Post-Implementation Success

The second level of the proposed model extends and respecifies the updated DeLone and McLean model (2003) to the GIS context via the following steps:

1) Replacing service quality dimension (that reflect the importance of service and support in successful e-commerce systems) with user quality. As GIS is not solely technical in nature, adding user quality dimension as a part of the process of producing geographic information is very important. GIS user quality is considered an important human factor in a successful GIS as mentioned in many GIS researches [62];

2) According to [15], researchers need to clearly and carefully define the stakeholders and the context in which Net Benefits are to be measured. Therefore, we split the net benefits dimension to net benefits to individuals, organization, and society;

3) DeLone and McLean IS success models (1992, 2003) are built upon the taxonomy developed by both Shannon and Weaver (1949) [64] and Mason (1978) [65], which considered that a message in a communication system can be measured at different levels including: the production level, the product level, and the influence level. The production level is the accuracy and efficiency of the system which produces the information. The product level is the success of the information in conveying the intended meaning. The influence level is the effect of the information on the receiver. The influence level is presented as a hierarchy of events taking place at the receiving end of an information system; these events are receipt of the information; influence of the information on the receiver; and the influence of the information on the performance of the system. The concept of levels of output from communication theory demonstrates the serial (process) nature of information (*i.e.*, a form of communication). The IS creates information which is communicated to the recipient who is then influenced (or not) by the information. In this sense, information flows through a series of stages

from its production through its use or consumption to its influence on individual and/or organizational performance [6]. Therefore, to reflect the interdependent and the process nature of the success dimensions that was suggested by DeLone and Mclean in their models (1992, 2003), we believe that system quality (which measures the success of the production level) and information quality (which measures the success of the product level) should not happen in parallel as DeLone and McLean (1992, 2003) did in their models. Instead, information quality should come after the system quality dimension to reflect the interdependent and the process nature of the success dimensions;

4) To avoid model complexity and to reflect the cross-sectional nature of this study, the feedback links from net benefits to both use and user satisfaction in the updated DeLone and McLean model (2003) were excluded.

The success dimensions of the proposed GIS success model are system quality, user quality, information quality, information use, user satisfaction, net benefits to individuals, net benefits to organization, and net benefits to society as shown in **Figure 4**. The arrows between success dimensions represent temporal (process) and causal relationships. The following section will discuss the GIS success dimensions.

The second level of the proposed model is concerned with measuring the GIS success after incorporating and adapting the GIS into the organization's operations. The GIS success should be measured after a wide spread of use to allow the members of the organization to arrive at informed opinions about the success of their GIS [29]. Hence it is termed GIS post-implementation success.

Model Construct

In this paper, according to previous researches on GIS, measures of the proposed model have been determined (see **Table 2**).

- System quality: system quality dimension measures the success of the technical aspects of GIS. System quality has been represented in many GIS studies by functionality, response time, system reliability, user friendless, error recovery, database content.
- User quality: user quality dimension represents the quality of GIS users in terms of spatial abilities and self efficacy. In IS field, Bonner (1995) [66] revised the DeLone and McLean model and introduced user quality in terms of knowledge skills and abilities. His recognition of the people element was a welcome addition to the model. Also, the recognition of the importance of human factor in evaluating GIS performance was first initiated by [67], who stated that people, not the computerized equipment, make a GIS success or fail.
- Information quality: information quality dimension is the quality of information provided to the organization using GIS, in the form of maps, tables, charts, and reports. The information quality dimension measured by accuracy, completeness, ease of interpretation, relevancy, reliability, timeliness, and clarity.
- Information use: information use is a broad construct that is frequently used in measuring the utilization of IS. Information use dimension measures to what extent the GIS output is being used in the decision making process. Clapp *et al.* (1989) [18] mentioned that, the system can provide the capability to obtain all desired information, but for some reasons, the information is not used in the decision process whether private or public. In this case, the GIS will fail due to lack of utilization. Use can be based on objective measures such, number of functions used, frequency of access, and amount of connecting time [68]. Questions about who uses the system, levels of use, motivations for and voluntariness of use, and the purpose and nature of system use are also relevant [67].
- User satisfaction: this dimension measures GIS user's level of satisfaction with the system. User satisfaction was traditionally employed as the most common measure of IS success. The most widely used user satisfaction instruments are End User Computing Support (EUCS) [69] instrument and User Information Satisfaction (UIS) [70]. Both the EUCS and the UIS instruments contain items related to system quality, information quality, rather than only measuring overall user satisfaction with the system. Because of this, some researchers have chosen to parse out the various quality dimensions from these instruments, and use a single item to measure overall satisfaction with an IS [71].
- Net benefits to individuals: this dimension summarizes benefits that can be gained by users when using GIS such as enhanced decision making, time saving, increase the understanding and awareness of problems [72].
- Net benefits to organization: this dimension summarizes the benefits that organization derives from using GIS, which refers to efficiency and effectiveness criteria. Efficiency is the degree to how GIS operates with minimum waste, duplication, and expenditure of resources, and can be expressed as cost savings, cost avoidance, or productivity gains. Efficiency may also result in the generation of revenue. Effectiveness involves generating a product of better quality or accomplishing an intended purpose [67].

Table 2. Measures of the model constructs.

Success Dimension		Measurement Items	References
System Quality	Functionality	The GIS software contains all the features and functions required to perform the required tasks	[29] [61] [67] [70] [73] [74]
	Response Time	Hardware and operating system response time are acceptable	
	Reliability	Server downtime typically 8 hours or less per year	
		all failures (including server, network, and software) are less than 40 hours downtime per year	
	User Friendless	GIS software is user-friendly	
	Error Recovery	It is easy to recover from errors encountered while using GIS software	
	Database Content	The database content is secured	
		Data backup is maintained throughout the organization	
		The database content is regularly updated throughout the organization	
		The database contains accurate data	
		The database contains all needed data for related tasks	
User Quality	Self Efficacy	Comfort to use	[63] [75]
		Capable to do	
		Understand what to do	
		Confidence to use	
	Spatial Abilities	Spatial ability test	
Information Quality	Accuracy	the GIS provides the accurate information you need	[18] [29] [59] [61] [67] [69]
	Completeness	the GIS provides sufficient information	
	Ease of Interpretation	the information on the map is easy to understand	
	Relevancy	the information provided meet your needs regarding your questions or problems	
	Reliability	the GIS provides reliable information	
	Timeliness	the GIS provide up to date information	
	Clarity	the information on digital or hardcopy maps are clear	
User Satisfaction	Technology Satisfaction	you are pleased with the GIS	[63] [69]
		you like to use the GIS	
		you are willing to use the GIS	
	Overall Satisfaction	Overall, how would you rate your satisfaction with the GIS?	
Information Use		To what extent do you actually use the reports or the output generated by the GIS?	[76] [77]
		To what extent could you get along without the use of the GIS?	
		What is the level of importance of decisions affected by the generated information?	
Net Benefits to Individuals	Time saving	Using GIS save time required for making decisions	[61] [67] [68]
	Enhanced Decision Making	As a result of GIS, I am better able to set my priorities in decision making	
		GIS has improved the quality of decisions I make in this organization (decisions are more accurate and correct)	
		As a result of GIS, the speed at which I analyze decisions has increased	

Continued

	Understanding	GIS enhances the understanding of the problems	
	Awareness	GIS enables timely problem recognition	
Net Benefits to Organization	Efficiency	The GIS helps the organization save cost in information production and provision	[17] [61] [73] [74]
		The GIS increases the organization profitability	
		The GIS improves the organization's competitive position	
	Effectiveness	The GIS helps the organization to achieve its goal	
		The GIS enables a new range of output (maps, tables, lists, etc.)	
		The GIS provides the organization with better motivated workforce	
		The GIS improves information sharing and flows to management and between departments	
		The GIS reduces risk in the decision making process	
Net Benefits to Society	Social Justice	The GIS provide equal availability of information to citizens when needed and equal ease of access	[18] [19] [28] [29] [72] [79]
	Participation	The GIS enables participation by public in decision process (Enhancement of principles of a democratic society)	
		Using the GIS improves the standard of health and safety in the society	
	Quality of Life	Using the GIS increases the economic benefits to the society	
		The GIS provides better service to public/citizens	

- Net benefits to society: based on the study of "The impact of GIS technology" conducted by [67], "societal impact" dimension has been proposed as a further variable to the lists of six DeLone and McLean's IS success dimensions. Societal impact is important to be considered in the evaluation of GIS success because the ultimate goal of all technologies introduced in the public sector agencies is to benefit society. Many researchers reported the benefits of GIS on broad societal objectives such as, citizen-public sector interactions, individual integrity, economic benefits, distribution of wealth and fulfillment of human aspirations, and equity.

3.3. Model Hypotheses

The first level of the proposed GIS success model acts like a guide or a reference for GIS project managers to concentrate on the most critical success factors of GIS project diffusion. Although, establishing the CSFs of GIS project does not implicate that the whole project will automatically succeed, but it would be erroneously to neglect one of these CSFs.

The second level of the proposed GIS success model is a multidimensional construct, and the dimensions are interrelated. GIS are first implemented and incorporated within the organization and exhibit various degrees of system, and user quality. System quality and user quality affect the quality of the produced information. Managers/decision makers experienced the quality of information by using it for their works. Users and managers/decision makers are either satisfied or not satisfied with using the GIS. Finally, the use of information by managers/decision makers and the satisfaction of GIS users trigger influence on net benefits to individuals, organization and society.

The second level of the proposed model suggests that there can be positive influence between the GIS success dimension. Thus, we propose the following 11 hypotheses:

H1. System quality will positively affect Information quality;

H2. User quality will positively affect Information quality;

H3a. Information quality will positively affect user satisfaction;

H3b. Information quality will positively affect information use;

H4. Information use will positively affect User satisfaction;

H5a. Information use will positively affect net benefit to individuals;

H5b. Information use will positively affect net benefit to organization;

H5c. Information use will positively affect net benefit to society;

H6a. User satisfaction will positively affect net benefit to individuals;

H6b. User satisfaction will positively affect net benefit to organization;

H6c. User satisfaction will positively affect net benefit to society.

4. Research Design and Method

4.1. Measures of the Constructs

To ensure the content validity of the scales used in this study, We used the measurement items that were operationalized and tested in previous empirical GIS/IS studies and were found to have demonstrated good psychometric properties. The measuring items for each success dimension are summarized in **Table 2**.

4.2. Data Collection Procedure

The data used to test the model were obtained from a sample of experienced GIS users and managers. This study developed a questionnaire (see appendix A) using a five-point Likert scale (1 - 5) ranging from "strongly disagree" to "strongly agree." The questionnaire was sent to 350 GIS users and managers in different GIS organizations in Egypt and abroad to answer the questions by assessing their GIS. For each question, respondents were asked to circle the response which best described their level of agreement. In total, 252 samples were received with an effective ratio of 72%. Detailed descriptive statistics relating to the respondents' characteristics are shown in **Table 3**.

Table 3. Respondents characteristics.

Characteristics	Number	Percentage
Job Title		
Decision Maker	7	2.8%
Geologist	19	7.5%
Geophysicist	6	2.4%
GIS Specialist	134	53.2%
Technician	38	15.1%
GIS Managers	48	19%
Gender		
Female	153	60.7%
Male	99	39.3%
Age		
21 - 30	87	34.5%
31 - 40	59	23.4%
41 - 50	62	24.6%
Over 50	44	17.5%
Work Experience		
1 - 5 years	82	32.5%
6 - 10 years	68	27%
11 - 15 years	40	15.9%
16 - 20 years	24	9.5%
Over 20 years	38	15.1%
Education Level		
Diploma	39	15.5%
Bachelor	162	64.3%
Master	30	11.9%
PhD	21	8.3%

5. Analysis and Results

5.1. Reliability Analysis

Reliability refers to the consistency or stability of the questionnaire results. Fewer errors lead to a higher level of reliability. In other words, a better reliability measurement will result from the consistency and stability of results. The present study measured the questionnaire reliability and the consistency of the items using Cronbach's alpha (SPSS Version 20). Many scholars have suggested that a Cronbach's alpha coefficient exceeding the 0.7 threshold indicates a high level of consistency among the aspects; a Cronbach's alpha coefficient exceeding 0.9 indicates a much higher level of consistency among the aspects (e.g., [79]-[83]). The Cronbach's alpha coefficients of the eight construct are greater than or equal the recommended value of 0.8 (see **Table 4**).

5.2. Model Analysis

Structural Equation Modeling technique was used to assess the model fit and show empirical findings and hypotheses results using LISREL 8.8. Seven common model-fit measures were used to assess the model's overall goodness of fit: the ratio of X^2 to degrees-of-freedom (df), goodness-of-fit index (GFI), adjusted goodness-of-fit index (AGFI), normalized fit index (NFI), comparative fit index (CFI), root mean square residual (RMSR), and root mean square error of approximation (RMSEA). As shown in **Table 5**, all the model-fit indices exceeded their respective common acceptance levels suggested by previous research, thus demonstrating that the proposed model exhibited a fairly good fit with the data collected. Thus, we could proceed to examine the path coefficient of the structural model.

Properties of the causal path, including the path coefficient, t-values, and variance explained for each equation in the hypothesized model, are presented in **Figure 5**.

System quality and user quality had significant positive influences on information quality. Thus, H1 and H2 were supported. The influence of information quality on both user satisfaction and information use were also significant. H3a and H3b were supported. The influence of information use on user satisfaction was not significant. Thus, H4 was rejected. User satisfaction has no significant impact on net benefits to (individuals, organization, and society). H6a, H6b, and H6c were rejected. Information use has a significant positive influence on both net benefits to individuals and net benefits to society, but has no significant effect on net benefits to organization. H5a and H5b were supported, while H5c was rejected. The direct, indirect, and total effects of system quality, user quality, information quality, information use and user satisfaction on net benefits to (individuals, organization, and society) are summarized in **Table 6**.

With regard to the constructs explained in the variance (R^2), 89% of the variance in information quality was explained by system quality and user quality, while 20% of the variance in information use was explained by information quality. 87% of the variance in user satisfaction was explained by information quality and information use. The variance explained by information use and user satisfaction on net benefit to individuals is 34%, net benefit to organization is 8%, and net benefit to society is 13%.

6. Discussion and Conclusions

In this paper, by combining IS success models and previous studies, a model for measuring GIS success is pre-

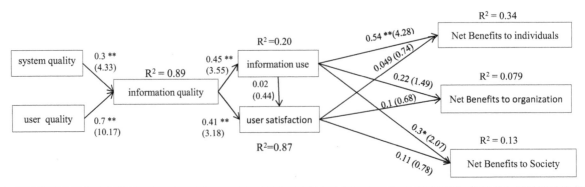

Figure 5. Hypotheses testing results (note: t-values for standardized path coefficient are described in parentheses). Statistically significant at $p < 0.05$; **Statistically significant at $p < 0.01$.

Table 4. Cronpach's alpha reliability analysis.

Construct (Cronbach Alpha)	Items	Total Relation of Fixed Item
System Quality (0.874)	SQ1	0.635
	SQ2	0.560
	SQ3	0.433
	SQ4	0.514
	SQ5	0.653
	SQ6	0.530
	SQ7	0.577
	SQ8	0.495
	SQ9	0.494
	SQ10	0.742
	SQ11	0.794
User Quality (0.93)	UQ1	0.835
	UQ2	0.886
	UQ3	0.919
	UQ4	0.740
Information Quality (0.91)	IQ1	0.831
	IQ2	0.761
	IQ3	0.725
	IQ4	0.781
	IQ5	0.842
	IQ6	0.543
	IQ7	0.671
Information Use (0.808)	IU1	0.908
	IU2	0.926
	IU3	0.831
User Satisfaction (0.95)	US1	0.870
	US2	0.833
	US3	0.431
	US4	0.705
Net Benefits to Individuals (0.997)	IND1	0.627
	IND2	0.686
	IND3	0.686
	IND4	0.686
	IND5	0.679
	IND6	0.668

Continued

	ORG1	0.627
	ORG2	0.632
	ORG3	0.617
	ORG4	0.620
Net Benefits to Organization (0.993)	ORG5	0.590
	ORG6	0.616
	ORG7	0.573
	ORG8	0.611
	SOC1	0.604
	SOC2	0.598
Net Benefits to Society (0.968)	SOC3	0.610
	SOC4	0.625
	SOC5	0.601

Table 5. Fit indices for structural model.

Fit Indices	Structural Model	Recommended Value
X^2/df	1.325	$<=3$
Goodness of Fit Index (**GFI**)	0.92	(0) to (1)
Adjusted Goodness of Fit Index (**AGFI**)	0.79	(0) to (1)
Root Mean Square Residual (**RMSR**)	0.051	$<=0.1$
Root Mean Square Error of Approximation (**RMSEA**)	0.081	$<=0.08$
Normed Fit Index (**NFI**)	0.95	$>=0.9$
Comparative Fit Index (**CFI**)	0.98	$>=0.9$
Relative Fit Index (**RFI**)	0.90	(0) to (1)

Table 6. The direct, indirect, and total effects.

	IQ			US			IU			IND			ORG			SOC		
	Direct	Indirect	Total Effect	Direct	Indirect	Total Effect	Direct	Indirect	Total Effect	Direct	Indirect	Total Effect	Direct	Indirect	Total Effect	Direct	Indirect	Total Effect
SQ	0.3	-	0.3	-	0.13	0.13	-	0.14	0.14	-	0.09	0.09	-	0.04	0.04	-	.06	0.06
UQ	0.7	-	0.7	-	-	-	0.31	-	0.31	-	0.25	0.25	-	0.16	0.16	-	0.19	0.19
IQ	-	-	-	0.41	-	-	0.45	-	-	-	0.29	0.29	-	0.14	0.14	-	0.18	0.18
US	-	-	-	-	-	-	-	-	-	0.05	-	0.05	0.1	-	0.1	0.11	-	0.11
IU	-	-	-	0.02	-	0.02	-	-	-	0.54	-	0.54	0.22	-	0.22	0.3	-	-

sented. The proposed GIS success model has comprehensive components. It integrates the CSFs (independent variables) with the outcome measures of GIS (dependent variables) into one model. The first level of the proposed model extends and respecifies the 3-D model of IS success by organizing the CSFs that have been discussed in the GIS literature, according to their occurrence in GIS diffusion stages. The first level of the proposed model can benefit organizations by focusing on the vital aspect of a successful GIS project. The second level of the proposed model is concerned with measuring the post-implementation success of GIS. The second level of the proposed model extends and respecifies the updated DeLone and McLean model (2003) in the context of GIS. Through the above analysis, 8 success dimensions, 49-items instrument were demonstrated to produce acceptable reliability estimates. The second level of the proposed model exhibited reasonable fit with the collected data. Six of the eleven hypotheses were found to be significant. The empirical results showed that the system quality and user quality had a significant positive influence on information quality. In addition, information quality had a positive influence on both information use and user satisfaction. It can be interpreted as a response to high system and user quality; a high information quality is produced which in turn causes satisfaction to GIS users and causes more use to these valuable information. The finding that information use did not have a significantly direct influence on user satisfaction was inconsistent with most prior IS studies. Thus, information use is necessary but not sufficient to cause user satisfaction. The results showed that information use had a significant influence on both net benefits to individuals and society, and that information use had no significantly positive effect on net benefits to organization. Also, user satisfaction had no significantly positive effect on net benefits to individuals, organization, and society. This explains why there are many different stakeholders whose satisfaction needs to be considered. The user satisfaction measures the technological satisfaction and the overall satisfaction of the GIS users (direct users) with their software, while the net benefits to individuals measure the satisfaction of the indirect users (like decision makers who make use of the technology by relying on other members of the organization) with GIS in enhancing the process of the decision making. The net benefits to organization are directed to GIS managers to measure the efficiency and the effectiveness of GIS in their organizations, while net benefits to society are directed to citizens to measure the contribution of GIS to societal well being especially in governmental organizations. This may explain why there is no direct causal relationship among these constructs. For example, the GIS users may be satisfied by using their GIS, but for some reasons the organization cannot achieve benefits due to administrative or economic problems. Therefore, user satisfaction should precede net benefits dimensions, but it is not sufficient to cause them.

This study is regarded as the first step in the long term research agenda of the researcher to develop and improve a model for measuring the GIS success. Therefore, the validity of a GIS success model cannot be truly established on the basis of a single study. Thus, caution needs to be taken when generalizing these findings. Validation of measurement requires the assessment of measurement properties over a variety of samples in similar and different contexts.

References

[1] Douglas, B. (2008) Achieving Business Success with GIS. John Wiley & Sons, London. http://dx.doi.org/10.1002/9780470985595

[2] Worboys, M. and Duckham, M. (2004) GIS: A Computing Perspective. CRC Press, Florida.

[3] Longley, P., Goodchild, M., Maguire, D. and Rhind, D. (2011) Geographic Information Systems and Science. John Wiley & Sons, Hoboken.

[4] Heywood, I., Corneliues, S. and Carver, S. (2011) An Introduction to Geographical Information Systems. Pearson, London.

[5] Roldan, J. and Leal, A. (2003) Executive Information Systems in Spain: A Study of Current Practices and Comparative Znalysis. In: Mora, M., Forgionne, G. and Gupta, J., Eds., *Decision Making Support Systems: Achievements, Trends, and Challenges for the New Decade*, Idea Group Publishing, London, 272-286. http://dx.doi.org/10.4018/978-1-59140-045-5.ch018

[6] DeLone, W.H. and McLean, E.R. (1992) Information Systems Success: The Quest for the Dependent Variable. *Journal of Information Systems Research*, **3**, 60-95. http://dx.doi.org/10.1287/isre.3.1.60

[7] Seddon, P.B., Staples, D.S., Patnayakuni, R. and Bowtell, M.J. (1999) Dimensions of Information Systems Success. *Journal of Communications of the Association for Information Systems*, **2**, 1-61.

[8] Ballantine, J., Bonner, M., Levy, M., Martin, A., Munro, I. and Powell, P. (1996) The 3-D Model of Information Systems Success: The Search of the Dependent Variable Continuous. *Journal of Information Resources Management*, **9**, 5-14. http://dx.doi.org/10.4018/irmj.1996100101

[9] DeLone, W.H. and McLean, E.R. (2003) The DeLone and McLean Model of Information Systems Success: A Ten-Year Update. *Journal of Management Information Systems*, **19**, 9-30.

[10] Seddon, P.B. (1997) A Respecification and Extension of the DeLone and McLean Model of IS Success. *Journal of In-*

formation Systems Research, **8**, 240-253. http://dx.doi.org/10.1287/isre.8.3.240

[11] Kaiser, M.G. and Ahlemann, F. (2010) Measuring Project Management Information Systems Success—Towards a Conceptual Model and Survey Instrument. *Proceedings of the 18th European Conference on Information Systems*, Pretoria, 1 January 2010, 1-12.

[12] Rai, A., Lang, S.S. and Welker, R.B. (2002) Assessing the Validity of IS Success Models: An Empirical Test and Theoretical Analysis. *Journal of Information Systems Research*, **13**, 50-69. http://dx.doi.org/10.1287/isre.13.1.50.96

[13] Seddon, P.B. and Kiew, M.Y. (1994) A Partial Test and Development of the DeLone and McLean model of IS Success. *Australasian Journal of Information Systems*, **4**, 90-109.

[14] Wu, J.H. and Wang, Y.M. (2006) Measuring KMS Success: A Respecification of the DeLone and McLean's Model. *Journal of Information & Management*, **43**, 728-739. http://dx.doi.org/10.1016/j.im.2006.05.002

[15] DeLone, W.H. and McLean, E.R. (2004) Measuring e-Commerce Success: Applying the DeLone and McLean Information Systems Success Model. *International Journal of Electronic Commerce*, **9**, 31-47.

[16] Antenucci, J.C., Brown, K., Crosswell, P.L., Kevany, M.J. and Archer, H. (1991) Geographic Information Systems: A Guide to the Technology. Springer, New York.

[17] Calkins, W. and Obermeyer, N. (1991) Taxonomy for Surveying the Use and Value of Geographical Information. *International Journal of Geographical Information Systems*, **5**, 341-351. http://dx.doi.org/10.1080/02693799108927859

[18] Clapp, J., McLaughlin, J., Sullivan, J. and Vonderohe, A. (1989) Toward a Method for the Evaluation of Multipurpose Land Information Systems. *Journal of the URISA*, **1**, 39-45.

[19] Tulloch, D.L. (1999) Theoretical Model of Multipurpose Land Information Systems Development. *Transaction in GIS*, **3**, 259-283. http://dx.doi.org/10.1111/1467-9671.00021

[20] Rockart, J.F. (1979) Chief Executives Define Their Own Data Needs. *Harvard Business Review*, **57**, 81-93.

[21] Bullen, C.V. and Rockart, J.F. (1981) A Primer on Critical Success Factors. Massachusetts Institute of Technology, Sloan School of Management, Massachusetts, USA.

[22] Yeoh, W., Koronios, A. and Goa, J. (2006) Critical Success Factors for the Implementation of Business Intelligence System in Engineering Asset Management Organizations. *Decision Support for Global Enterprise*, **2**, 34-51.

[23] Campbell, H. and Masser, I. (1995) GIS and Organizations. Taylor and Francis, London.

[24] Anderson, C.S. (1996) GIS Development Process: A Framework for Considering the Initiation, Acquisition, and Incorporation of GIS Technology. *Journal of URISA*, **8**, 10-26.

[25] Campbell, H. (1992) Organizational Issues and the Implementation of GIS in Massachusetts and Vermont: Some Lessons for the United Kingdom. *Environmental and Planning B: Planning and Design*, **19**, 85-95. http://dx.doi.org/10.1068/b190085

[26] Onsrud, H.J. and Pinto, J.K. (1993) Evaluating Correlates of GIS Adoption Success and the Decision Process of GIS Acquisition. *Journal of URISA*, **5**, 18-39.

[27] Rogers, E.M. (1995) Diffusion of Innovations. The Free Press, New York.

[28] Huxhold, W.E. and Levinsohn, A.G. (1995) Managing Geographic Information System Projects. Oxford University Press, New York.

[29] Obermeyer, N.J. and Pinto, J.K. (2008) Managing Geographic Information Systems. Guilford Press, New York.

[30] Biehl, M. (2007) Success Factors for Implementing Global Information Systems. *Communications of The ACM*, **50**, 53-58. http://dx.doi.org/10.1145/1188913.1188917

[31] Kohan, M., Yusoff, W. and Asadi, A. (2011) KSFs in GIS Adoption during Crisis Management. *Proceedings of the International Conference on Sociality and Economics Development*, Kuala Lumpur, 15 May 2011, 331-335.

[32] Nasirin, S. and Birks, D. (1998) Geographical Information Systems (GIS) Success Factors amongst UK Food Retailers: Comparisons between Market Leaders and Followers. *Proceedings of the 10th Annual Colloquium of the Spatial Information Research Center*, Dunedin, 16 November 1998, 235-242.

[33] Alwaraqi, G. and Zahary, A. (2012) Critical Factors of GIS Projects Failure in Yemeni Governmental Agencies. *Proceedings of the 13th International Arab Conference on Information Technology*, Jordan, 10-13 December 2012, 53-65.

[34] Birks, D., Nasirin, S. and Zailani, S. (2003) Factors Influencing GIS Project Implementation Failure in the UK Retailing Industry. *International Journal of Information Management*, **23**, 73-82. http://dx.doi.org/10.1016/S0268-4012(02)00036-1

[35] Openshaw, S., Cross, A., Charlton, M. and Brunsdon, C. (1990) Lessons Learnt from a Post Mortem of Failed GIS. *Proceedings of the 2nd National Conference and Exhibition of the AGI*, Brighton, 1 January 1990, 231-235.

[36] Clarke, A. (1991) GIS Specification, Evaluation and Implementation. In: Goodchild, M.F., Maguire, D.J. and Rhind,

D.W., Eds., *Geographic Information Systems and Sciences*, John Wiley & Sons, Hoboken, 477-448.

[37] Somers, R. (1998) Developing GIS Management Strategies for an Organization. *Journal of Housing Research*, **9**, 157-178.

[38] Sieber, R.E. (2000) GIS Implementation in the Grassroots. *Journal of URISA*, **12**, 15-29.

[39] Velibeyoglu, K. (2004) Institutional Use of Information Technologies in City Planning Agencies: Implications from Turkish Metropolitan Municipalities. PhD Dissertation, Izmir Institute of Technology, Izmir.

[40] Croswell, P.L. (1991) Obstacles to GIS Implementation and Guidelines to Increase the Opportunities for Success. *Journal of URISA*, **3**, 43-56.

[41] Man, W.H. and Toorn, W.H. (2002) Culture and the Adoption and Use of GIS within Organizations. *International Journal of Applied Earth Observation and Geoinformation*, **4**, 51-63.
http://dx.doi.org/10.1016/S0303-2434(02)00013-2

[42] Rumor, M. (1993) The Use of Geographic Information Technology in the City of Paova. In: Masser, I. and Onsrud, H.J., Eds., *Diffusion and Use of Geographic Information Technologies*, Kluwer Academic Publishers, Dordrecht, 229-243.

[43] Campbell, H. (1994) How Effective Are GIS in Practice? A Case Study of British Local Government. *International Journal of Geographical Information Systems*, **8**, 309-325. http://dx.doi.org/10.1080/02693799408902002

[44] Cavric, B. (2002) Human and Organizational Aspects of GIS Development in Botswana. *Proceedings of the 6th GSDI Conference*, Pudabest, 10 October 2002, 1-18.

[45] Hussain, M. and Johar, F. (2010) The Socio-Technical Factors in the Use of GIS at the Planning Departments of the Kuala Lumpur City Hall Planning Malaysia. *Journal of the Malaysian Institute of Planners*, **8**, 69-103.

[46] Dekolo, S.O. (2001) Implementing GIS for Land Use Planning and Management in Lagos State. *Proceedings of the URISA Conference*, California, 20 October 2001, 173-182.

[47] Higgs, G., Smith, D.P., Myles, I. and Gould, M.I. (2005) Findings from a Survey on GIS Use in the UK National Health Service: Organizational Challenges and Opportunities. *Health Policy*, **72**, 105-117.
http://dx.doi.org/10.1016/j.healthpol.2004.06.011

[48] Lamer, N. and Disera D. (1997) System Success Factors and the Top Ten Reasons AM/FM/GIS Projects Falter. *Proceedings of the Geospatial Information and Technology Association Conference*, Nashvill, 401-404.

[49] Davis, J. (1999) Six Keys to Gaining Executive Commitment to GIS. *Proceedings of the URISA Conference*, Chicago, 11 September 1990, 566-570.

[50] Demens, M. (2005) Fundamentals of Geographic Information Systems. John Wiley & Son, Hoboken.

[51] Gallaher, D. (1999) Three Leading Killers of GIS. *Proceedings of the URISA Conference*, Chicago, 21 August 1999, 585-586.

[52] Klosterman, R.E. (1995) The Appropriateness of Geographic Information Systems for Regional Planning in the Developing World. *Computers, Environment and Urban Systems*, **19**, l-13.
http://dx.doi.org/10.1016/0198-9715(94)00028-X

[53] Nasirin, S. and Birks, D. (2003) DSS Implementation in the UK Retail Organizations: A GIS Perspective. *Journal of Information and Management*, **40**, 325-336.

[54] Otawa, T. (2004) Benefits and Obstacles of GIS Implementation: Recent Perceptual Shifts and Implications for City and Regional Planning Organizations. *Proceedings of the Geospatial Information and Technology Association Conference*, Seattle, April 2004, 1-9.

[55] Somers, R. (2001) Quick Guide to GIS Implementation and Management. URISA, Chicago.

[56] Nedovic-Budic, Z. and Godschalk, D. (1996) Human Factors in Adoption of Geographic Information Systems: A Local Government Case Study. *Public Administration Review*, **56**, 554-67. http://dx.doi.org/10.2307/977254

[57] Taleai, M., Mansourian, A. and Sharifi, A. (2009) Surveying General Prospects and Challenges of GIS Implementation in Developing Countries: A SWOT-AHP Approach. *Journal of Geographical Systems*, **11**, 291-310.
http://dx.doi.org/10.1007/s10109-009-0089-5

[58] Beck, D. (1997) Avoid Top 10 Mistakes GIS Project Mistakes. *GIS World*, **10**, 48-51.

[59] Brown, M. (1996) An Empirical Assessment of the Hurdles to Geographic Information System Success in Local Government. *State and Local Government Review*, **28**, 193-204.

[60] Gillespie, S.R. (2000) An Empirical Approach to Estimating GIS Benefits. *Journal of URISA*, **12**, 7-13.

[61] Aldaijy, E.Y. (2004) The Multidimensional Measurements of Geographic Information Systems (GIS) Effectiveness in Crisis Management. *Proceedings of the* 2004 *Command and Control Research and Technology Symposium*, Monterey,

June 2004, 1-22.

[62] Bone, T. and Johnson, D. (2007) Human Factors in GIS Use: A Review and Suggestions for Research. *Proceedings of the Information Systems Education Conference*, Pittsburgh, 3 November 2007, 1-15.

[63] Jarupathirun, S. and Zahedi, F. (2003) Exploring the Influence of Perceptual Factors in the Success of Web-Based Spatial DSS. *Decision Support Systems*, **43**, 933-951. http://dx.doi.org/10.1016/j.dss.2005.05.024

[64] Shannon, C. and Weaver, W. (1949) The Mathematical Theory of Communication. University of Illinois Press, Chicago.

[65] Mason, R. (1978) Measuring Information Output: A Communication Systems Approach. *Information and Management*, **1**, 219-234. http://dx.doi.org/10.1016/0378-7206(78)90028-9

[66] Bonner, M. (1995) DeLone and McLean's Model for Judging Information Systems Success—A Retrospective Application in Manufacturing. *Proceedings of the European Conference on IT Investment*, Henley Management College, Berkshire, 11 July1995, 218-226.

[67] Nedovic-Budic, Z. (1999) Evaluating the Effects of GIS Technology: Review of Methods. *Journal of Planning Literature*, **13**, 284-295. http://dx.doi.org/10.1177/08854129922092405

[68] Chein, S. and Tsaur, S. (2007) Investigating the Success of ERP Systems: Case Studies in Three Taiwanese High-Tech Industries. *Journal of Computers in Industry*, **58**, 783-793. http://dx.doi.org/10.1016/j.compind.2007.02.001

[69] Doll, W.J. and Torkzadeh, G. (1988) The Measurement of End-User Computing Satisfaction. *MIS Quarterly*, **12**, 259-272. http://dx.doi.org/10.2307/248851

[70] Ives, B., Olson, M. and Baroudi, J.J. (1983) The Measurement of User Information Satisfaction. *Communications of the ACM*, **26**, 785-793. http://dx.doi.org/10.1145/358413.358430

[71] Petter, S., DeLone, W.H. and McLean, E.R. (2008) Measuring Information Systems Success: Models, Dimensions, Measures, and Interrelationships. *European Journal of Information Systems*, **17**, 236-263. http://dx.doi.org/10.1057/ejis.2008.15

[72] Akingbade, A., Navarra, D. and Georgiadou, Y. (2009) A 10 Years Review and Classification of the Geographic Information Systems Impact Literature 1998-2008. *Nordic Journal of Surveying and Real Estate Research*, **4**, 84-116.

[73] Witkowski, M.S., Rich, P.M. and Keating, G.N. (2007) Metrics of Success for Enterprise GIS. *Journal of Map and Geography Libraries*, **4**, 59-82. http://dx.doi.org/10.1300/J230v04n01_04

[74] Nedovic-Budic, Z., Pinto, J.K. and Budhathoki, N.R. (2008) SDI Effectiveness from the User Perspective. In: Crompvoets, J., Rajabifard, A., Loenen, B. and Fernández, T.D., Eds., *A Multi-View Framework to Assess Spatial Data Infrastructures*, The Melbourne University Press, Melbourne, 273-303.

[75] Lee, J. and Bednarz, R.S. (2012) Components of Spatial Thinking: Evidence from a Spatial Thinking Ability Test. *Journal of Geography*, **111**, 15-26. http://dx.doi.org/10.1080/00221341.2011.583262

[76] Franz, C.R. and Robey, D. (1986) Organizational Context, User Involvement, and the Usefulness of Information Systems. *Decision Sciences*, **17**, 329-356. http://dx.doi.org/10.1111/j.1540-5915.1986.tb00230.x

[77] Zwart, P. (1991) Some Indicators to Measure the Impact of Land Information Systems in Decision Making. *Proceedings of the URISA Conference,* Washington DC, 11 August 1991, 68-79.

[78] Sanders, G.L. and Coutney, J.F. (1985) A Field Study of Organizational Factors Influencing DSS Success. *MIS Quarterly*, **9**, 77-93. http://dx.doi.org/10.2307/249275

[79] Ganapati, S. (2010) Using GIS to Increase Citizen Engagement. IBM Center for the Business of Government, Washington DC.

[80] Chang, L.M., Chang, S.L., Ho, C.T., Yen, D.C. and Chiang, M.C. (2011) Effects of IS Characteristics on e-Business Success Factors of Small- and Medium-Sized Enterprises. *Computers in Human Behavior*, **27**, 2129-2140. http://dx.doi.org/10.1016/j.chb.2011.06.007

[81] Cooper, D.R. and Schindler, P.S. (2001) Business Research Methods. McGraw-Hill Higher Education, London.

[82] Hair, J.F., Anderson, R.E., Tatham, R.L. and Black, W.C. (2006) Multivariate Data Analysis with Reading. Prentice Hall, Upper Saddle River.

[83] Wang, H.C. and Chiu, Y.F. (2009) Assessing e-Learning 2.0 System Success. *Computers and Education*, **57**, 1790-1800. http://dx.doi.org/10.1016/j.compedu.2011.03.009

Appendix A. Survey Items Used in This Study

1) System quality

SQ1: The GIS software contains all the features and functions required to perform the required tasks

SQ2: The hardware and operating system response time are acceptable

SQ3: The server downtime typically 8 hours or less per year

SQ4: All failures (including server, network, and software) are less than 40 hours downtime per year

SQ5: The GIS software is user-friendly

SQ6: It is easy to recover from errors encountered while using GIS software

SQ7: The database content is secured

SQ8: Data backup is maintained throughout the organization regularly

SQ9: The database content is regularly updated throughout the organization

SQ10: The database contains accurate data

SQ11: The database contains all the needed data for related tasks

2) User quality

UQ1: You feel comfort while using the GIS

UQ2: You are capable to do the required task

UQ3: You understand what you do

UQ4: You feel confidence while using the GIS software

UQ5: Spatial ability test (20 questions developed by Lee and Bednarz (2012))

3) Information quality

IQ1: The GIS provides the accurate information you need

IQ2: The GIS provides sufficient information

IQ3: The information on the map is easy to understand

IQ4: The information provided meet your needs regarding your questions or problems

IQ5: The GIS provides reliable information

IQ6: The GIS provide up to date information

IQ7: The information on digital or hardcopy maps are clear

4) User satisfaction

US1: You are pleased with the GIS

US2: You like to use the GIS

US3: You are willing to use the GIS

US4: Overall, how would you rate your satisfaction with the GIS?

5) Information use

IU1: To what extent do you actually use the reports or the output generated by the GIS?

IU2: To what extent could you get along without the use of the GIS?

IU3: What is the level of importance of decisions affected by the generated information?

6) Net benefits to individuals

IND1: Using GIS save time required for making decisions

IND2: As a result of GIS, I am better able to set my priorities in decision making

IND3: GIS has improved the quality of decisions I make in this organization

IND4: As a result of GIS, the speed at which I analyze decisions has increased

IND5: GIS enables timely problem recognition

IND6: GIS enhances the understanding of the problems

7) Net benefits to organization

ORG1: The GIS helps the organization save cost in information production and provision

ORG2: The GIS increases the organization profitability

ORG3: The GIS improves the organization's competitive position

ORG4: The GIS helps the organization to achieve its goal

ORG5: The GIS enables a new range of output (maps, tables, lists, etc.)

ORG6: The GIS provides the organization with better motivated workforce

ORG7: The GIS improves information sharing and flows to management and between departments.

ORG8: The GIS reduces risk in the decision making process

8) Net benefits to society

SOC1: The GIS provide equal availability of information to citizens when needed and equal ease of access

SOC2: The GIS enables participation by public in decision process (enhancement of principles of a democratic society)

SOC3: Using the GIS improves the standard of health and safety in the society

SOC4: Using the GIS increases the economic benefits to the society

SOC5: The GIS provides better service to public/citizens

10

Potential Erosion Risk Calculation Using Remote Sensing and GIS in Oued El Maleh Watershed, Morocco

Hicham Lahlaoi[1], Hassan Rhinane[1], Atika Hilali[1], Said Lahssini[2], Loubna Khalile[1]

[1]Geosciences Laboratory, Faculty of Sciences Ain Chock, Hassan II University, Casablanca, Morocco
[2]National School of Forestry Engineering, Salé, Morocco
Email: Lahlaoi.hicham@gmail.com

Abstract

Oued El Maleh watershed is considered the largest ocean basin of the Chaouia-Ouardigha region in Morocco. Severe flooding occurred in 1996, 2001 and 2002 in the watershed. Thus, significant economic and human damage has been caused. The floods of Mohammedia city, located in the outlet of the watershed, were due to the silting of the Oued El Maleh dam which has lost its ability to retain water. This work, therefore, aims to assess soil losses by water erosion in the Oued El Maleh watershed through modeling main factors involved in water erosion. The methodology used is based on the use of the universal soil loss equation (USLE). The model includes the following factors: soil erodibility, the inclination of slopes, the rainfall erosivity, vegetation cover and erosion control practices. The aggressiveness of rainfall was calculated for a number of stations bordering the study area and interpolated across the watershed using geostatistical model. Soil erodibility was extracted from soil map and soil survey. The effect of topography was approached by combining the degree of slope and slope length using a digital elevation model (ASTER) and ArcHydrology extension (ArcGIS). The vegetation cover was derived from Landsat image ETM through the supervised classification method. The index of erosion control practices was approached by field visits. All factors have been measured and integrated into a geographic information system which enabled us to spatialize the degree of sediment production at the watershed scale in a synthetic map. The annual soil loss is 8.21 t/ha/yr and the soil loss classification shows that surfaces affected by high erosion are equivalent to 10% of the watershed. Furthermore, this map is available to support land managers policy makers in the process of decision making related to soil conservation, infrastructure and citizens' property protection.

Keywords

Watershed Oued El Maleh, Erosion, USLE, Geographic Information System

1. Introduction

Water erosion is a dynamic process of detaching, transporting and depositing soil particles under the effect of the kinetic energy of water. Soil loss causes adverse influences of widespread with different intensities depending on the environment biophysical characteristics and threats human sustainability [1]. The effects of this phenomenon are not limited only to the reduction of agricultural land productivity [2], but they also affect the quantity and quality of available water by accelerating the rate of siltation of reservoirs [3] and reducing the production of electricity [4] [5].

In turkey the erosion rate reaches an average which varies from 500 to 600 million ton/year [6]. In Syria, the soil erosion risk map published by FAO, UNEP and UNESCO in 1980 indicated losses values ranging between 50 and 200 t/ha/yr. In Lebanon, the estimates reported figures ranging from 50 to 70 t/ha/yr in the mountains of Lebanon and Anti-Lebanon (FAO, 1986). In Tunisia, 45% of the land area is threatened by erosion [7]. In Algeria, 45% of Tellian areas, equivalent to 12 million hectares, are threatened [8].

In Morocco, 40% of land is affected by water erosion [9]. In some parts of the Rif in northern Morocco, erosion rates sometimes reach 30 to 60 t/ha/yr [10] [11] and get to 2000 t/km^2/yr for the central and western Rif [12]. In the Middle and High Atlas the annual averages ranged from 500 to 1000 t/km^2/year and from 1000 to 2000 t/km^2/yr in the Pre-Rif and the Mediterranean border. Therefore, dams lose their water storage initial capacity due to their siltation which is estimated at 0.5% by year [13]. The largest Moroccan dams receive each year approximately 50 million tonnes of sediment [14], which affects their storage capacity and brings about an annual loss of almost 300 million Dirhams [15].

In the watershed of Oued El Maleh scale, the erosion is manifested in different forms and it's accentuated by the degradation of Mdakra and Achach forests. In fact, the local population subsistence depends on forest products (wood, firewood, grazing…) and agriculture [16]. This anthropogenic pressure on natural resources increases the land vulnerability to erosion.

In this area, where floods are important, the construction of the recent dam Tamesna located upstream needs protection to combat siltation. The objective of this work is to study the vulnerability of watershed erosion through the use of empirical model USLE [17]. It is organized into three sections: 1) materials and methods with the presentation of the study area and the main factors involved in the model (these were evaluated by field investigations associated with the use of remote sensing and/or derived by the use of GIS); 2) section results relating the implementation of the model and 3) discussions and conclusions section.

2. Materials and Methods

2.1. Presentation of the Study Area

Oued El Maleh watershed is located in the regions of Chaouia Ouardiga. It extends on provinces of Mohammedia, Settat, Benslimane and Khouribga, between 32°90 N - 33°76 N lalitude and 06°60 E - 07°50 E longitude. It's bounded on the North-east by Oued Nfifikh watershed and by oceanic bassins of Chaouia on the south-west (**Figure 1**). It covers an approximate area of 2577 km^2. It is the catchment area drained by the Oued Mellah and its affluents Oued hassar, oued Zamra, oued Laatach and whose outlet is located at the city of Mohammedia (**Figure 2**).

The bioclimate is semi-arid with a rainfall annual average of 320 mm. the altitude is ranging from 0 to 962 meters. Geologically, the area is still characterized by large outcrops of Triassic series of red clay. Several soils types were identified, mainly vertisoils, fersialitics, calcimagnesics and poorly evolved soils.

2.2. Data Used

The data used in this study include land use, digital elevation model (DEM), rainfall data and soil data. The land use was extracted from Landsat image Enhanced Thematic Mapper (ETM) acquired on December 16, 2014 with a spatial resolution of 30 m and projected in Universal Transverse Mercator (UTM). The digital elevation model (DEM) was extracted for the study area from ASTER Global Digital Elevation Model (ASTER GDEM) with spatial resolution of 30 m. The rainfall data for 30 years were collected from Hydraulic Basin Agency of Bouregreg and Chaouia over 12 climatic stations. The pedology data were provided by the National Institute for Agricultural Research of Settat.

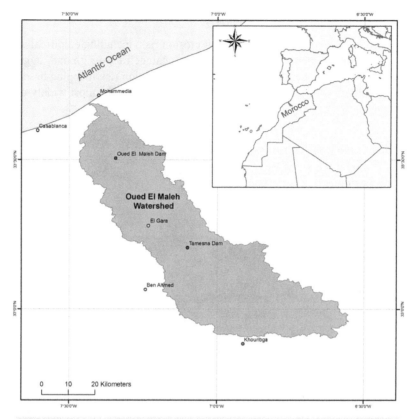

Figure 1. Situation of Oued El Maleh watershed.

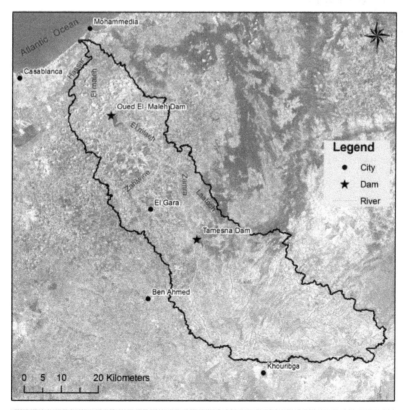

Figure 2. Hydrographic network of Oued El Maleh watershed.

2.3. Used Model and Methodology

Different approaches have been used to assess the soil erosion risk, including empirical erosion models [18] [19], a ranking method based on selected indicators such as percentage of bare ground, aggregate stability, organic carbon, percentage clay, and bulk density [20] and qualitative erosion risk mapping based on the combination of five factors namely geology, soil, relief, climate and vegetation [21]. The most widely used model is the USLE [22] expressed as follows:

$$A = R \times K \times LS \times C \times P \tag{1}$$

where: A is the soil loss (t·ha^{-1}·yr^{-1}). R is the rainfall erosivity factor (MJ·mm·ha^{-1}·h^{-1}·yr^{-1}). K is the soil erodibility factor (t·h·MJ^{-1}·mm^{-1}). LS is the slope steepness and slope length factor (dimensionless). C is the vegetation cover factor (dimensionless). P is the conservation practice factor (dimensionless).

2.3.1. Rainfall Erosivity Factor (R)
The original equation of (R) uses the kinetic energy of the rain and requires measurements of rainfall intensity (Wischmeier & Smith, 1978):

$$R = K \ Ec \ I30 \tag{2}$$

This direct method of Wischmeier and Smith can only be applied in areas equipped with autographic recorders. An alternative formula developed by Wischmeier and Smith (1978) and modified by Arnoldus (1980) [23] involves only annual and monthly precipitation to determine the R factor.

$$R = 1.735 \times 10^{1.5 \cdot \log \Sigma \left(pi^2 / P \right) - 0.8188} \tag{3}$$

The annual and monthly precipitation was recovered from 12 weather stations for 30 years. R values were calculated for the selected stations (**Table 1**). They were interpolated over the whole watershed using geostatistic model (kriging).

2.3.2. Soil Erodibility Factor (K)
The soil erodibility factor K represents the susceptibility of soil particles to be detached. It's related to the integrated effects of rainfall, runoff, and infiltration on soil loss, accounting for the influences of soil properties on soil loss during storm events on upland areas [24]. K values were derived using the wischmeier nomograph [25] for survey soil analysis (**Table 3**) and the soil map. The K is estimated through the following experimental equation.

Table 1. Annual and monthly precipitations data.

Station	J	F	M	A	M	J	J	A	S	O	N	D	P
AînKheil	43	63	61	44	14	3	0	1	12	31	46	48	367
Ahmed Ben Ali	52	46	31	33	6	1	0	1	3	24	34	50	281
Al Gara	59	53	51	38	17	3	1	0	6	33	53	71	383
Barrage O. Maleh	55	46	42	38	18	4	2	2	6	26	48	73	359
Ben Ahmed	62	47	50	39	17	3	1	2	7	31	54	66	376
Bouznika	55	47	55	31	21	6	1	0	6	31	67	77	397
Berrchid	62	50	49	34	16	3	1	1	8	31	57	67	376
Bir Baiz	32	28	23	32	14	8	0	6	7	21	33	48	375
El khatouate	50	60	42	45	21	6	13	2	7	29	30	36	340
Mohammedia	61	53	54	46	20	6	0	1	7	34	71	77	431
Oued Zem	50	42	43	37	18	7	4	2	7	26	48	48	375
Settat	58	54	45	34	14	2	0	1	5	36	53	69	400

Source: Hydraulic Basin Agency of Bouregreg and Chaouia.

$$100K = 2.1M^{1.14} \cdot 10^{-4} (12 - OM) + 3.25(S - 2) + 2.5(P - 3) \tag{4}$$

where: K is the soil erodibility factor (t·ha·MJ^{-1}·mm^{-1}), M: (% silt + %fine sand) × (100 − %clay), OM is the % of organic matter, S is the soil structure code, P is the permeability code.

The k index was calculated for each soil type by integrating soil analysis values in the study area and generalized through the soil map.

2.3.3. Topographic Factor (LS)

The (LS) factor reflects the combined effect of slope length and slope steepness on erosion. The empirical equation developed by Wischmeier & Smith is done by following formula.

$$LS = \left(\frac{L}{22.13}\right)^m \times \left(0.065 + 0.045 \cdot S + 0.0065 \cdot S^2\right) \tag{5}$$

where: L is the slope length in meters, S is the angle of slope in percent, m is a constant dependent on the value of the slope gradient: 0.5 if the slope angle is greater than 5%, 0.4 on slopes of 3% to 5%, 0.3 on slopes of 1 to 3%, and 0.2 on slopes less than 1%.

To implement LS factor in Arc GIS, the below formula of Bizwuerk *et al.* (2008) was used [26].

$$LS = \left(FA \times \frac{CS}{22.13}\right)^m \times \left(0.065 + 0.045 \cdot S + 0.0065 \cdot S^2\right) \tag{6}$$

where FA is the flow-accumulation and CS is the cell size.

The flow-accumulation was derived from a MNE ASTER, using Arc-hydrology in spatial analyst extension, according the following algorithm (**Figure 3**).

2.3.4. Cover Management Factor (C)

Vegetation plays an important role in protecting soil against erosion. The vegetation canopy intercepts the rainfall, increases the infiltration and reduces the rainfall kinetic energy.

The C factor is defined as the ratio of soil loss from land cropped under specific conditions to the corresponding loss from clean-tilled, continuous fallow (Wischmeier and Smith, 1978). Currently, due to the variety of land cover patterns with spatial and temporal variations, satellite remote sensing data sets were used for the assessment of C factor [27]. The C factor has a close linkage to land use types [28]. Ground cover were collected in sample plots with a GPS and reported in Landsat image ETM, the land use at non-sampled location were determined through a supervised classification technique [29].

The major land use types of OM watershed are: agiculture (33.3%), pasture (32.5%), and forest ecosystems (16.5%). The values assigned to each land use are recorded in **Table 2**; it is ranging between 0.05 for dense Forest and 1 for badlands.

2.3.5. Support Practices Factor (P)

The P factor explains human intervention in creating erosion control practices that conserve soil and reduce surface runoff [30]. These practices include contouring, strip-cropping, terracing, strips, etc. [31].

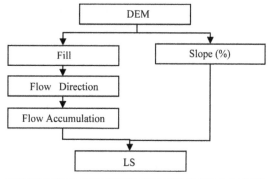

Figure 3. LS factor calculation algorithm.

Table 2. C factor for land use.

Land Use	C Index
Dense Forest	0.05
Open Forest	0.10
Plantation	0.13
Matorral	0.20
Pasture	0.30
Agriculture	0.60
Bad Lands	1.00

Table 3. Physic properties of soil types.

Soil type	Silt (%)	Clay (%)	Fine Sand (%)	Organic Matter (%)	Structure code	Permeability (%)
Calcimanesic Soils	15.1	19.46	41.86	2.81	3	22.34
Isohumic Soils	14.03	37.38	15.00	3.28	2	25.33
Poorly Evolved Soils	18.68	25.84	18.68	1.47	1	13.81
Vertisoils	19.29	53.29	18.07	2.17	4	18.28
Fersialitic Soils	7.98	16.95	56.25	1.47	2	13.82
Hydromorphic Soils	13.03	11.85	41.12	1.53	2	14.18
Raw Mineral Soils	18.68	25.84	27.47	1.47	2	13.82
Brown Soils	16.54	15.44	20.85	2.70	3	21.65
Brown Soils	16.54	15.44	20.85	2.70	3	21.65

Source: National Institute for Agricultural Research of Settat.

The values of P-factor ranges from 0 to 1, in which the highest value is assigned to areas with no conservation practices; the minimum values correspond to built-up-land and plantation area with strip and contour cropping. The lower the P value, the more effective the conservation practices. For Oued El Maleh watershed, except two perimeters namely Seffoud and Bouhrar, were we found brenched plantation, there is no significant support practice.

3. Results

3.1. Evaluation of R, K, LS, C, and P factors

The R values were based on the formula modified by Arnoldus using annual and monthly precipitations, the data were provided from 12 stations for 30 years. R values oscillate between 62.59 to 104.63 $MJ \cdot mm \cdot ha^{-1} \cdot h^{-1} \cdot yr^{-1}$ with an average of 79.06 $MJ \cdot mm \cdot ha^{-1} \cdot h^{-1} \cdot yr^{-1}$ (**Figure 4**). We observe a weak spatial variation of R which increases following the South-East to North-West direction.

At the watershed level, the erodibility index K is between 0.07 and 0.39 $t \cdot h \cdot MJ^{-1} \cdot mm^{-1}$ (**Figure 5**). The vertisoils, isohumic soils and brown soils are little erodible whereas raw mineral soils, poorly evolved soils and calcimanesic soils, which are mainly encountered in the basin are moderately erodible.

On the other Hand, fersialitic soils and hydromorphic soils have high erodibility values, reflecting their high susceptibility to erosion.

Figure 6 shows that the majority of the study area has LS Values less than 1.5. Some specific areas with a big steep slope, such as along the river have LS values greater than 4.5.

The values assigned to land cover range between 0.05 for dense forest and 1 for badlands. 0.6 was attributed to agriculture which is the dominant land use (**Figure 7**).

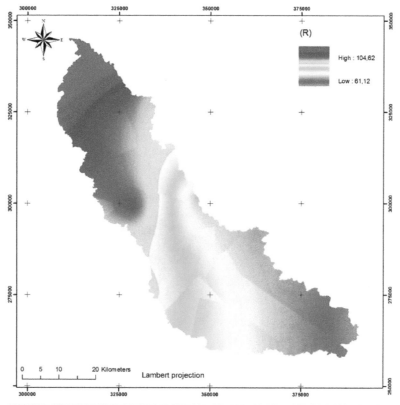

Figure 4. Erosivity factor (R).

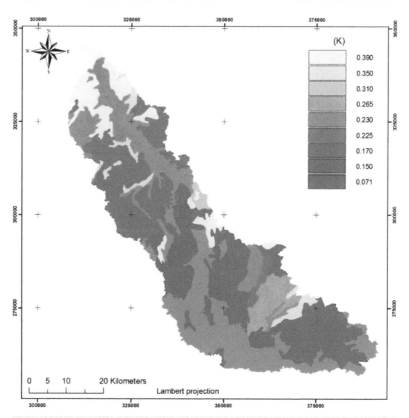

Figure 5. Erodibility factor (R).

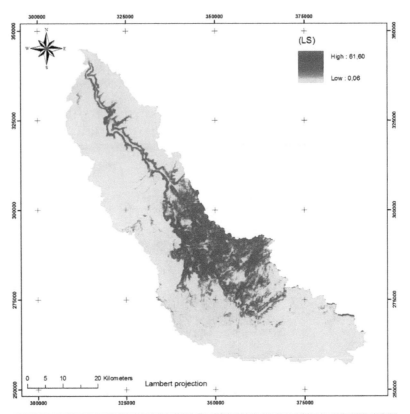

Figure 6. Topography factor (LS).

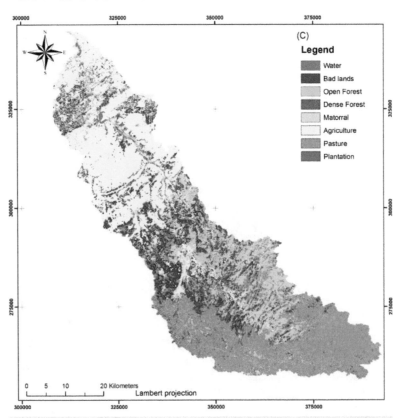

Figure 7. Land cover factor.

For both Seffoud and Bouhrar perimeters, we assigned the P factor value of 0.5 and of 1 for all other areas of the watershed.

3.2. Evaluation Soil Loss

The empirical USLE model was implemented in ArcGIS [32] using map algebra on raster layers on different indices involved in the model. All the factors were overlaid to generate the soil loss map (**Figure 8**). The maximum and minimum losses are respectively about 0.02 t·ha^{-1}·yr^{-1} and 501.40 t·ha^{-1}·yr^{-1}. The average value per hectare is 8.21 t·ha^{-1}·yr^{-1}.

Sadiki and al (2009) reported a soil classification system according to their susceptibility to erosion [33]. Indeed, soils can tolerate a significant agricultural production with soil loss not exceeding 7 t·ha^{-1}·yr^{-1}. Beyond 7 t·ha^{-1}·yr^{-1}, soil losses are important and can compromise agricultural production. Between 20 and 30 t·ha^{-1}·yr^{-1}, erosion is considered high and is considered too high from the threshold 35 t·ha^{-1}·yr^{-1}.

The analysis of **Table 4** shows that 73% of the area is exposed to law erosion, which explains why agriculture is the dominant practice in the study area.

Sub-catchment and soil loss maps were integrated to compute erosion for different sub-catchment (**Figure 9**). According to the calculated Values, Laatach sub-catchment records the highest average of soil loss (**Table 5**). However, it's Zamra sub-catchment which more contributes to sediment production. This could be explained by its big area.

4. Conclusions

The watershed of Oued El Maleh suffered severe climatic aggressiveness with an average of 79.07 MJ·mm·ha^{-1}·h^{-1}·yr^{-1}. The weighted average value of k for the watershed is 0.21 with a median of 0.23, confirming the susceptibility of soils to erosion in this watershed.

The average LS value is 1.26. According to the grading standards of Manrique (1988) related to LS factor, we can say that most of the land in the watershed belongs to the low risk class (0 - 2 units). The highest value of LS coincides with rough areas.

Table 4. Classification of soil loss in watershed Oued El Maleh.

Soil Loss (t/ha/Yr)	Intensity	Area (Km^2)	Area %
0 - 7	Weak	1891.00	73.38
7 - 20	Moderate	440.67	17.10
20 - 30	Strong	87.36	3.39
30 - 35	Very Strong	26.80	1.04
>35	Extremely Strong	130.65	5.07
Total		2576.48	100

Table 5. Prioritization of sub-catchment based on soil loss assessment.

Sub Catchment	Min (t/ha/yr)	Max (t/ha/yr)	Mean (t/ha/yr)	Sum (t/yr)	Priority N°
Zamra	0.02	464.22	8.60	1060287.35	1
Laatach	0.04	501.39	18.44	557208.25	2
Oued El Maleh Dam's Upstream	0.05	442.12	8.35	191112.64	3
Oued El Maleh Dam's Downstream	0.07	417.83	8.23	155600.64	4
Zahiwine	0.02	121.02	3.46	99758.39	5
Hassar	0.05	114.63	2.21	74757.30	6
Total	0.02	501.39	8.21	2138724.56	

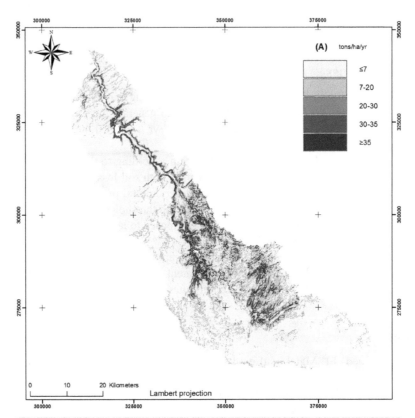

Figure 8. Annual soil loss in watershed Oued El Maleh.

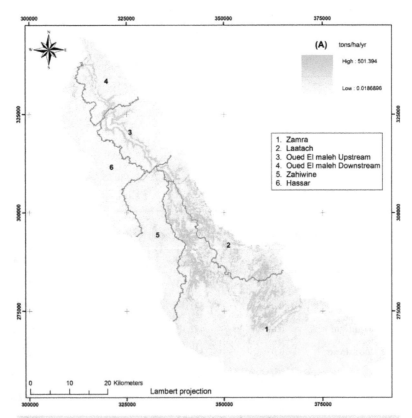

Figure 9. Annual soil loss in Oued El Maleh sub-catchments.

Potential Erosion Risk Calculation Using Remote Sensing and GIS in Oued El Maleh...

137

The role of vegetation is crucial. The average C value in the watershed is 0.5. This reflects the modes of space use characterized by both a dominance of agriculture that less protects the soil as well as the scarcity of the forest domain which well protects the soil both by its root system and its canopy.

The rate of erosion is generally considered moderate in most of the watershed with spatial variability. Moreover, it appears from the analysis of the potential erosion map that the majority of produced sediment comes from Zamra and Laatach sub-catchment whilst the minimum sediment produced is generated in Hassar subcatchment.

In general, USLE is used to estimate average annual soil loss, the use of remote sensing and GIS allows us to spatialize the potential risk of erosion however; the uncertainties regarding data may introduce uncertainties about soil loss estimates.

Although the USLE has been criticized for its lack of applications in areas different from its development as well as for giving erroneous results, it is still regarded as best generally available model that has been virtually tested in all environments of the world. Bouqdaoui (2007) has synthesized them [34]. Sadiki and al (2009) concluded that the model underestimates the real soil loss which seems normal due the fact that the model doesn't take into account other forms of erosion.

On the whole, this work has approached the problem of erosion in the watershed of Oued El Maleh, which has experienced several floods. At its end, the use of GIS and remote sensing has facilitated the identification of the erosive potential of the watershed and identify sub-basins that contribute most in the production of sediment. This is a support decision making and may help guide managers for selecting priorities to be addressed in antierosion management areas with the purpose to preserve human lives and infrastructure.

References

[1]   Lal, R. (1998) Soil Erosion Impact on Agronomic Productivity and Environment Quality: Critical Reviews. *Plant Sciences*, **17**, 319-464.

[2]   Parveen, R. and Kumar, U. (2012) Integrated Approach of Universal Soil Loss Equation (USLE) and Geographical Information System (GIS) for Soil Loss Risk Assessment in Upper South Koel Basin, Jharkhand. *Journal of Geographic Information System*, **4**, 588-596. http://dx.doi.org/10.4236/jgis.2012.46061

[3]   Apusiga Adongo, T., Kugbe, J.X. and Gbedzi, V.D. (2014) Siltation of the Reservoir of Vea Irrigation Dam in the Bongo District of the Upper East Region, Ghana. *International Journal of Science and Technology*, **4**, 2224-3577.

[4]   Sharma, R., Sahai, B. and Karale, R.L. (1985) Identification of Erosion Prone Areas in Part of the Ukai Catchment. *Proceedings of the Sixth Asian Conference of the Remote Sensing*, Hyderabad, India, November 1985, 121-126.

[5]   Bunyasi, M.M., Onywere, S.M. and Kigomo, M.K. (2013) Sustainable Catchment Management: Assessment of Sedimentation of Masinga Reservoir and Its Implication on the Dam's Hydropower Generation Capacity. *International Journal of Humanities and Social Science*, **9**, 166-179.

[6]   Celik, I., Aydin, M. and Yazici, U. (1996) A Review of the Erosion Control Studies During the Republic Period in Turkey. In: Kapur, S., Akça, E., Eswaran, H., Kelling, G., Vita-Finzi, Mermut, A.R. and Ocal, A.D., Eds., 1st *International Conference on Land Degradation*, Adana, Turkey, 10-14 June 1996, 175-180.

[7]   Boussema, M.R. (1996) Système d'Information pour la Conservation et la Gestion des Ressources Naturelles. *Colloque international sur le rôle des technologies de télécommunications et de l'information en matière de protection de l'environnement*, Tunis, 17-19 avril 1996.

[8]   Chebbani, R., Djilli, K. and Roose, E. (1999) Etude des Risques d'Erosion dans le Bassin Versant de l'Isser, Algérie. *Bulletin Réseau Erosion*, **19**, 85-95.

[9]   Chevalier, J.J. Pouliot, J., Thomson, K. and Boussema, M.R. (1995) Systèmes d'Aide à la Planification Pour la Conservation des Eaux et des Sols (Tunisie). Systèmes d'Information Géographique Utilisant les Données de Télédétection. *Actes du colloque scientifique international*, Hammamet, Tunisie, 1-2 Novembre 1994, 4-12.

[10] Lahlou, A. (1977) Specific Degradation of Watershed in Morocco. Report n° 1000, Ministry of Equipment and National Promotion, Water Direction, Exploitation Division, Water Management Service, Rabat, 1977.

[11] Ait Fora, A. (1995) Modélisation Spatiale de l'Erosion Hydrique dans un Bassin Versant du Rif Marocain: Validation de l'Approche Géomatique par la Sédimentologie, les Traceurs Radio-Actifs et la Susceptibilité Magnétique des Sédiments. Ph.D. Thesis, Sherbrooke University, Quebec.

[12] Mhirit, O. and Benchekroun, F. (2006) Les Ecosystèmes Forestiers Marocains: Situation, Enjeux et Perspectives Pour 2025. Rapport sur le Développement Humain au Maroc, Rabat.

[13] Tahri, M., Merzouk, A., Lamb, H.F. and Maxted, R.W. (1993) Etude de l'Erosion Hydrique dans le Plateau d'Imelchil

dans le Haut Atlas Central. Utilisation d'un SIG. *Geo Observateur*, **3**, 51-60.

[14] Merzouki, T. (1992) Diagnostic de l'envasement des grands barrages marocains. *Revue marocaine du génie civil*, **38**, 46-50.

[15] High Commission for Water, Forest and Combating Desertification (1996) National Watershed Management Plan.

[16] Chalrhami, H. (2014) Analysis of Anthropogenic Impacts on Achach Forest Degradation on Central Plateau, Morocco. *International Journal of Latest Research in Science and Technology*, **3**, 176-180.

[17] Wischmeier, W.H. and Smith, D.D. (1960) An Universal Soil Loss Estimating Equation to Guide Conservation Farm Planning. *7th International Congress of Soil Sciences*, 418-425.

[18] Boggs, G., Devonport, C., Evans, K. and Puig, P. (2001) GIS-Based Rapid Assessment of Erosion Risk in a Small Catchment in the Wet/Dry Tropics of Australia. *Land Degradation & Development*, **12**, 417-434. http://dx.doi.org/10.1002/ldr.457

[19] Cerri, C.E.P., Dematte, J.A.M., Ballester, M.V.R., Martinelli, L.A., Victoria, R.L. and Roose, E. (2001) GIS Erosion Risk Assessment of the Piracicaba River Basin, Southeastern Brazil. *Mapping Sciences and Remote Sensing*, **38**, 157-171.

[20] Shakesby, R.A., Coelho, C.O.A., Schnabel, S., Keizer, J.J., Clarke, M.A., Contador, J.F.L., Walsh, R.P.D., Fereira, A.J.D. and Doerr, S.H. (2002) A Ranking Methodology for Assessing Relative Erosion Risk and Its Application to Dehesas and Montados in Spain and Portugal. *Land Degradation & Development*, **13**, 129-140. http://dx.doi.org/10.1002/ldr.488

[21] Vrieling, A., Sterk, G. and Beaulieu, N. (2002) Erosion Risk Mapping: A Methodological Case Study in the Colombian Eastern Plains. *Journal of Soil and Water Conservation*, **57**, 158-163.

[22] Wischmeier, W.H. and Smith, D.D. (1978) Predicting Rainfall Erosion Losses: A Guide to Conservation Planning. Science, US Department of Agriculture Handbook, No. 537, Washington DC.

[23] Arnoldus, H.M.J. (1980) An Approximation of the Rainfall Factor in the Universal Soil Loss Équation. In: De Boodt, M. and Gabriels, D., Eds., *Assessment of Erosion*, John Wiley and Sons, New York, 127-132.

[24] Renard, K.G., Foster, G.R., Weesies, G.A., McCool, D.K. and Yoder, D.C. (1997) Predicting Soil Erosion by Water: A Guide to Conservation Planning with the Revised Universal Soil Loss Equation (RUSLE). US Department of Agriculture Handbook, No. 703, Washington DC, 1-251.

[25] Wischmeier, W.H. and Smith, D.D. (1978) Predicting Rainfall Erosion Losses: A Guide to Conservation Planning. Science, US Department of Agriculture Handbook, No. 537, Washington DC.

[26] Bizuwerk, A., Taddese, G. and Getahun, Y. (2008) Application of GIS for Modeling Soil Loss Rate in Awash Basin, Ethiopia. International Livestock Research Institute, Addis Ababa, Ethiopia

[27] Karydas, C.G., Sekuloska, T. and Silleos, G.N. (2009) Quantification and Site-Specification of the Support Practice Factor When Mapping Soil Erosion Risk Associated with Olive Plantations in the Mediterranean Island of Crete. *Environmental Monitoring and Assessment*, **149**, 19-28. http://dx.doi.org/10.1007/s10661-008-0179-8

[28] Prasannakumar, V., Vijith, H., Abinod, S. and Geetha, N. (2012) Estimation of Soil Erosion Risk within a Small Mountainous Sub-Watershed in Kerala, India, Using Revised Universal Soil Loss Equation (RUSLE) and Geo-Information Technology. *Geoscience Frontiers*, **3**, 209-215. http://dx.doi.org/10.1016/j.gsf.2011.11.003

[29] Fadil, A., Rhinane, H., Kaoukaya, A., Kharchaf, Y. and Bachir, O.A. (2011) Hydrologic Modeling of the Bouregreg Watershed (Morocco) Using GIS and SWAT Model. *Journal of Geographic Information System*, **3**, 279-289. http://dx.doi.org/10.4236/jgis.2011.34024

[30] Renschler, C.S., Mannaerts, C. and Diekkruger, B. (1999) Evaluating Spatial and Temporal Variability in Soil Erosion Risk—Rainfall Erosivity and Soil Loss Ratios in Andalusia, Spain. *CATENA*, **34**, 209-225. http://dx.doi.org/10.1016/S0341-8162(98)00117-9

[31] Angima, S.D., Stott, D.E., O'Neill, M.K., Ong, C.K. and Weesies, G.A. (2003) Soil Erosion Prediction Using RUSLE for Central Kenyan Highland Conditions. *Agriculture, Ecosystems and Environment*, **97**, 295-308. http://dx.doi.org/10.1016/S0167-8809(03)00011-2

[32] ESRI (2008) Arc GIS Software of ESRI.

[33] Sadiki, A., Faleh, A., Zezera, J.L. and Mastas, H. (2009) Quantification de l'Erosion en Nappes dans le Bassin Versant de l'Oued Sahla-Rif Central Maroc. *Cahiers Géographiques*, **6**, 59.

[34] Bouqdaoui, K. (2007) Approche Méthodologique de l'Evaluation du Risque Potentiel d'Erosion des Sols du Bassin Versant d'Oued Srou à l'Aide de la Télédétection et du SIG. PhD Thesis, University of Mohamed V, Rabat.

11

Mashing up Geographic Information for Emergency Response—An Earthquake Prototype

Shawn Dias, Chaowei Yang, Anthony Stefanidis, Mathew Rice

Department of Geography and GeoInformation Science, George Mason University, Fairfax, VA, USA
Email: sdias@gmu.edu, cyang3@gmu.edu, astefani@gmu.edu, rice@gmu.edu

Abstract

Important information pertaining to emergencies and responses to the emergencies is often distributed across numerous Internet sites. In the event of a disaster like an earthquake, rapid access to such information is critical. At such moments the general public usually has a hard time navigating through numerous sites to retrieve and integrate information, and this may severely affect our capability to make critical decisions in a timely manner. Common earthquake mashups often lack relevant information like locations of first responders and routing to important facilities (e.g. hospitals and fire stations) which could save important time and lives. To address the challenges, we developed an Earthquake Information Mashup prototype. This prototype demonstrates a mashup approach to providing a Web visualization of real-time earthquake monitoring and complementary information, such as traffic conditions, the location of important facilities and routing to them. It also offers users the ability to communicate local condition. Users are thus able to better integrate information from various near real-time sources, obtain better situational awareness, and make smarter informed critical decisions.

Keywords

CyberGIS, Natural Hazards, Social Media, Performance, CloudGIS

1. Introduction

Natural disasters like earthquakes often strike unexpectedly and leave the public incapable of dealing with the situation. Most earthquake studies [1] [2] have shown that people often panic in the direct aftermath of the event.

It is during this time that quick and informed decisions need to be made. At that time information fragments become available across multiple Internet sites [3]. For the public, coming to grips with the situation and making an effort to compose all this information are equivalent to putting together an extremely complex jigsaw puzzle with pieces scattered all over a room, and some of them even missing. Issues like format variation, differences in projections, outdated content and lack of synchronization present often insurmountable challenges for the general public.

On August 2011, an earthquake struck Northern Virginia with a magnitude of 5.8 and an epicenter five miles southwest of Mineral Springs and left its footprints all across the east coast. Given the rarity of earthquakes in the US east coast, this one generates high levels of anxiety, and the general public rushed to the Internet for information. While most were downloading information, many also contributed by posting event-related narrative to a variety of platforms. Most notable, twitter was used heavily by social media users who acted as hybrid seismographs to report the event [4] [5]. This represents an alternate type of volunteered content, also referred to as ambient geographical information (AGI) [6], which complements and expands volunteered geographical information (VGI). This crowd-contributed content has been proven very valuable during natural disasters such as the Haitian Earthquake [7], the Red River Floods [8] and the Santa Barbara Wild Fires [9].

When faced with disasters like earthquakes which affect large geographical areas, real time sensors can be incorporated in a GIS environment to raise situation awareness. With the evolution of Web 2.0 there has been an increase in the number of sensors connected to the Web providing real time data that feed the Internet with vast volumes of information. Government agencies like USGS and NOAA map disasters and distribute real time wildfire and earthquake data to the general public [10]. For emergency management, GIS technologies can be used in mitigation, preparedness, response and recovery [11].

Earthquakes occur due to the sudden release of energy in the earth's crust. They are evidence of tectonic activity and occur when tectonic stress is released by movement of rocks along a fault [12]. Observers and scientists gather information about earthquakes such as intensity levels by noting the damage and residents experiences during the earthquake. People's Web postings have also been used to assess the earthquake impact [13]. A GIS is particularly appropriate for emergency response as it communicates a better view of the event impact, and the distribution of the resources that are available to respond to the post-event conditions. It allows end-users to select data necessary to analyze, to identify spatial patterns, to solve resource allocation problems, and to make good spatial decisions [14]. For example after a 7.0 magnitude earthquake, GIS was used to answer questions like where to search for victims and where rescue teams should be assigned [15].

In order to integrate distributed Web content we make use of WebGIS solutions that employ languages such as JavaEE (Java Enterprise Edition), Servlet, JSP technologies for developing applications, and operate on a Web server such as a Tomcat server. Data between the client and server can be exchanged using different formats, popular ones include XML (Extensible Markup Language) and JSON (JavaScript Object Notation). Web services which allow other programs to call are usually responded in XML or JSON format which allows programs to parse. Two popular Web services include SOAP and REST: 1) SOAP (Simple Object Access Protocol) uses a structured or encapsulated XML for exchanging information. It is difficult to construct and parse hence HTTP is preferred; 2) REST (Representational State Transfer) was introduced and can be implemented based on HTTP where the client sends all parameters in the request URL. Clients usually consume Web services like geo browsers such as Google Maps, Bing Maps to build customized application [16].

With the evolution of Web 2.0 there has been an increase in the number of sensors connected to the Web providing real time data sets. WebGIS offers functionalities like visualization of large amounts of GIS data, and analysis of this is possible [17]. A WebGIS tool could be used to visualize and analyze complex variables and models such as climate data [18] [19]. Web based mapping projects such as Wikimapia and Open Street Map allow users to collect geospatial Information [20]. WebGIS also provides a platform for distributing geospatial information, Websites such as ArcGIS.com allow users to download data and use its Web services. The availability of information through WebGIS and open data format enables us to integrate multiple data sources to provide comprehensive spatial decision support information through mashups. A mashup in Web applications refers to a practical approach that brings together existing elements to create refined functionalities in a seamless interoperable fashion [21]. Just as a music DJ mixes more than one song similarly users can mix data from more than one Website [22] [23].

Mashups can be done at either server side or browser side: [21] Server-side mashups were used in all mashups prior to 2005. They have powerful software and hardware, and involve complex server side programming. They are also labor intensive due to the time required for development and deployment. Browser-side mashups function through processing on the client computer. They use relevant Web services and technologies such as JavaScript, AJAX and XML. Some sites provide data via JavaScript, which is used by the browser side mashups. Other mapping companies provide maps and services via a JavaScript API. These APIs are easy to use and do not involve complex programming. Browsers can make use of Web services in the form of RESTful and SOAP services.

Making use of the vast amount of unstructured content on the Internet can be difficult [17]. Content sent to mashups may be of two types: with or without an API. Mashups which make use of API, REST and SOAP services are relatively simple to construct and are often created by amateurs [17] [23]. The ease with which they can be created has prompted the development of a large variety of mashups from simple mashups involving simple dynamic data to large mashups using big data. There has been a lot of interest in crisis mashups as it is often used to describe neogeographic practices [24]. During crises, timely decisions are essential. An automated visualization system can provide enhanced situational awareness to decision makers after or during a disaster and could help them make better spatial decisions [25] [26]. Visualizing this combined information from a mashup not only enables decision makers to see things they never could see before, but also helps them make quick intelligent decisions. Therefore, we investigated how a mashup could be used to integrate spatial information for earthquake emergency response.

2. Related Work

2.1. Rich Internet Application

Since the 1990s, Web interfaces have been developed with features such as multimedia and animation, improving user satisfaction and resulting in a user experience similar to desktop application [27]. Ajax applications make use of asynchronous communication behind the scenes to provide smooth interactions while the graphical user interfaces are being used. Browser side APIs make the Web application fast, fun and easy to use. Rich Internet Application technologies can be used to help users obtain required geospatial resources easily [28]. They allow users to choose different options, all in one page. Tasks are accomplished on demand thus eliminating the need for going through multiple steps and reloading of pages. Rich Internet technologies are also beneficial to developers as they are easy to build thus reducing developing time and costs.

2.2. Mashup

Each mashup is unique in its own way, however, most current earthquake mashups focus on communicating the location, magnitude and time of the earthquake. There are many mashups present with relevant functionalities like

- The AEGIS, Advanced Emergency GIS, monitors and maps the location and status of emergencies, locates victims and emergency-response personnel, and tracks other factors that can impact emergency response [21].
- The USGS [29] has the Earthquake Hazard Program to reduce earthquake losses in the US [30]. The USGS mashup offers viewing and monitoring of earthquakes that have occurred in the past 24 hours and are greater than 2.5 in magnitude.
- The Hungarian National Association of Radio Distress Signaling and Info communications [31] runs an Emergency and Disaster Information Service (EDIS) as an important resource for travelers to get information about hazards like forest fires, earthquakes, tropical storms, and disease outbreaks.
- IRIS Earthquake Browser [32] enables the public to find earthquakes in any region of the globe and then import this information into the GEON Integrated Data Viewer (IDV) where the hypocenters may be visualized in three dimensions.
- Rapid Earthquake View [33] allows views and monitoring of recent earthquakes and access to seismograms from seismograph stations around the world.
- The Live Earthquake Map [5] shows earthquakes in the last 24 hours. This mashup uses a Google Map geographic frame for the base map and its feed Web services are obtained from three different sources.
- ESRI has a similar Public Information Map for Earthquakes [34] which shows the earthquakes for the last 90 days, population in the affected area and a shake map.

2.3. Grid Clustering

Plotting thousands of markers on the map is computationally challenging and thus may lead to a degraded experience. Grid-based clustering [35] was used to restrict the markers rendered on the map. It worked by dividing the map into squares of a certain size and then grouping the markers into each grid square. This type of rich and engaging user experiences can not only increase the performance but can also give better user satisfaction.

Witnessing the need for integrating information for decision making after an earthquake and the maturity of relevant technologies, we propose a mashup mechanism and present a prototype to demonstrate its potential. Section 3 introduces the design architecture. Section 4 discusses the implementation. Section 5 demonstrates the system through several scenarios. Section 6 concludes and discusses future research.

3. System Design and Architecture

3.1. Framework

The system is designed with an architecture (**Figure 1**) includes application server, database, client, live feeds, VGI input and leveraging google maps.

- **Client side:** The application was developed within Web client, integrating CSS, JavaScript, and additional toolkits like Dojo for providing a better user experience. It makes use of Browser-side API such as Google Maps API which provides a geographic frame for the basemap, Google Traffic API which is used to obtain live traffic conditions, and Google Directions API which is used to provide routing to essential facilities within a twenty mile buffer. Google Maps consumes the RESTful Web services from USGS in GeoJSON Format (**Figure 2**). The user can also provide input by filling out a form describing the earthquake and reporting nearby events. When the user sends a URL request to the local Tomcat server it connects to Web services, receives a response and displays HTML. The browser interprets this HTML markup and displays it in a form the client can understand. The user can also select the first responder facilities like emergency operation centers, fire stations, police stations and medical care facilities he would like to view and routing to facilities within a 20 mile buffer. He can select the option to view the live traffic conditions around him. The user can also provide VGI inputs by means of a form describing the earthquake and reporting the events around him.
- **Server side:** Operations such as reading or writing to a database, parsing of data or Web scrapping need to be performed on the Tomcat server. These operations are complex and done on the server side through powerful server side programming languages such as Java. The live feeds are obtained from the USGS server.

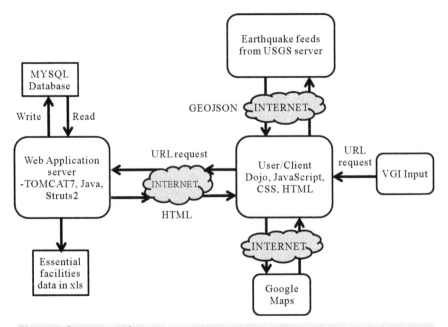

Figure 1. System overview.

USGS Server

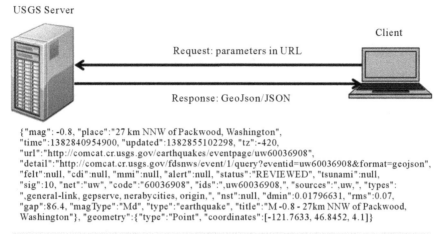

{"mag": -0.8, "place":"27 km NNW of Packwood, Washington",
"time":1382840954900, "updated":1382855102298, "tz":-420,
"url":"http://comcat.cr.usgs.gov/earthquakes/eventpage/uw60036908",
"detail":"http://comcat.cr.usgs.gov/fdsnws/event/1/query?eventid=uw60036908&format=geojson",
"felt":null, "cdi":null, "mmi":null, "alert":null, "status":"REVIEWED", "tsunami":null,
"sig":10, "net":"uw", "code":"60036908", "ids":",uw60036908,", "sources":",uw,", "types":
",general-link, gepserve, nerabycities, origin,", "nst":null, "dmin":0.01796631, "rms":0.07,
"gap":86.4, "magType":"Md", "type":"earthquake", "title":"M -0.8 - 27km NNW of Packwood,
Washington"}, "geometry":{"type":"Point", "coordinates":[-121.7633, 46.8452, 4.1]}

Figure 2. REST-style Web service, request in the form of a URL and response in JSON.

- **Database:** Data such as the VGI input and essential facilities information (e.g. emergency operation centers, fire station, police station, and medical care facilities) are stored in a MySQL database, and can be added and retrieved as desired.

3.2. Methodologies

Thick client architecture was employed, relying on the client rather than the server to perform most of the functions. This was made possible by using the Dojo toolkit [36] which is a rich Internet application allowing asynchronous communication between the browser and the server. There is fast interaction with the user since information on the page is updated without the need to reload the whole page [37]. The Dojo toolkit supports a rich user experience. There is also less pressure on the server since there are fewer round trips to the server. However it is critical to reduce the volume of data that have to be processed on the client to reduce delays.

RESTful Services are used for the Web application so that the client can send the request in the form of URL and receive the response in the form of JSON.

Model Viewer Controller (MVC) Architecture was used since it is the most currently used architecture to build Web based enterprise systems. MVC Architecture (**Figure 3**) was used to separate each module into three parts-the model, controller, and viewer for promoting the maintenance and separation of tasks to each component. It allows the division of work onto separate modules.

Marker Cluster: Many Web applications have latency problems, described commonly by end-users as "sluggishness". Various techniques (such as pyramids and hash indices, multithreading and caching) have been employed to improve the experience [38]. The Web application developed in this research project requires the display of the locations of a large number of essential facilities. Grid based clustering was used to reduce the location markers displayed on the map by iterating though the markers in the list that needs to be clustered and adding each one into the closest cluster if it is within a minimum square pixel bounds (**Figure 4**). This increased the performance of the application and caused less visual overload.

4. Prototype Implementation

4.1. Data Types and Sources

4.1.1. Google Maps
Google Maps [39] was used as a geographic framework of reference because of its powerful Traffic and Directions API makes available live traffic conditions. The only restriction is that Google Maps API allows 25,000 requests per day and charges users when the request number exceeds 25,000.

4.1.2. Earthquake Feeds
The live earthquake feeds were obtained from USGS (United States Geological Survey). GeoJSON Format was chosen from many formats [29] as it was smaller in size and easier to parse compared to XML. USGS provides

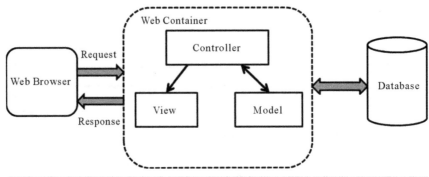

Figure 3. MVC Architecture.

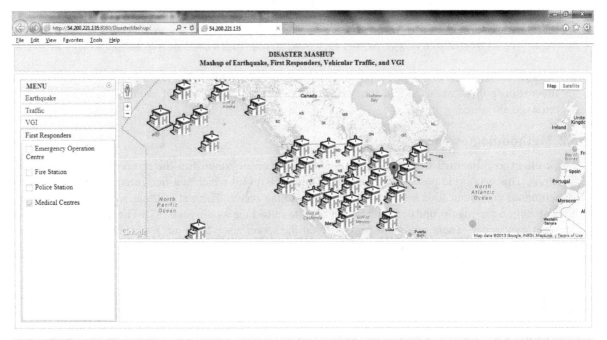

Figure 4. Grid based clustering.

information of all earthquakes greater than 2.5 magnitudes for the past week:
http://earthquake.usgs.gov/earthquakes/feed/geojsonp/2.5/week. The European Mediterranean Seismological Centre (EMSC) distributes data on earthquakes in the last 24 hours. Big earthquakes events from around the world are shown, but EMSC focuses on European countries with smaller events from across Europe, especially the Mediterranean Sea (Greece, Aegean Sea and Turkey) region. The Geo Forschungs Zentrum (GFZ) Potsdam provides data for large earthquakes in the last twenty four hours, and has global coverage.

4.1.3. Essential Facilities Data

Essential facilities are those that provide services to the community and should be functional after an earthquake. They include fire stations, police stations, emergency operation centers and medical care centers (which include hospitals and medical clinics).

Data for essential facilities at the national level is difficult to obtain. Local governments, such as US counties typically have a rich library of information. However, combining county level data for the entire United States is a large, complex, and difficult task. The data for the emergency operation centers, hospitals, fire stations and police stations for the entire United States can be obtained from the software HAZUS-MH [11] (**Figure 5**). HAZUS-MH is a GIS based natural disaster (earthquake, hurricane and flood) modeling tool developed by Federal Emergency Management Agency (FEMA) to provide data to strengthen preparedness and response capabilities by assessing risk and forecasting losses.

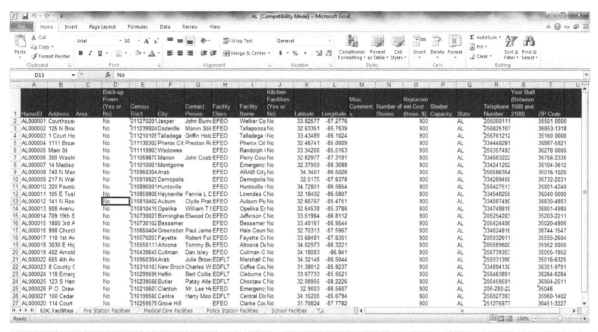

Figure 5. Screen shot of data obtained from HAZUS-MH for the state of Alabama.

The essential facilities are extremely important, because after a disaster these sites provide the public with assistance and are critical in dealing with emergencies and supporting the public. The medical care facilities data within the HAZUS-MH software come from the American Hospital Association (AHA) 2000 data, the emergency response, fire stations and police stations data come from Info USA Inc. 2001 data. The HAZUS-MH software has a tool called Comprehensive Data Management System (CDMS) which allows the inbuilt inventory to be exported.

This data exported from CDMS is by state. Each state excel file contained sheets which holds the different facilities information.

4.1.4. Live Traffic

Live traffic conditions, and routing to facilities can be obtained from Google Traffic API Web services [40] (**Figure 6**), providing a user with live traffic conditions to find the best route to his destination. The color codes indicate the speed of traffic compared to free flowing traffic conditions. The Google Traffic API is supported primarily in urban areas, and the data is not always highly accurate, from temporal and qualitative perspective, but can add useful information during a crisis. Cell phone companies are constantly monitoring location of smartphone users based on the strength of signal received at the cell towers. When smartphone users use Google Maps and enable their location option on their smartphones, they are enabling their GPS. With the triangulation method between the cell towers and the GPS the mobile phone can be tracked and traffic information can be obtained through this technique and through several other related techniques.

4.1.5. Routing to Essential Facilities

The routing to essential facilities was obtained using the Google Directions API Web services [41]. The Google Directions API is a service that receives direction requests and returns routing results. In the request three mandatory fields need to be sent:

1) *Origin*, specifies the start location from which directions need to be calculated. In the Earthquake Information Mashup tool, the user's current location is used.

2) *Destination*, specifies the end location to which directions need to be calculated. In the Earthquake Information Mashup tool, one of the essential facilities selected by the user will be used.

3) *travelMode*, this parameter specifies what mode of transportation will be used. This value can include walking, bicycling, transit and driving which is the default. In the Earthquake Information Mashup tool, driving was used.

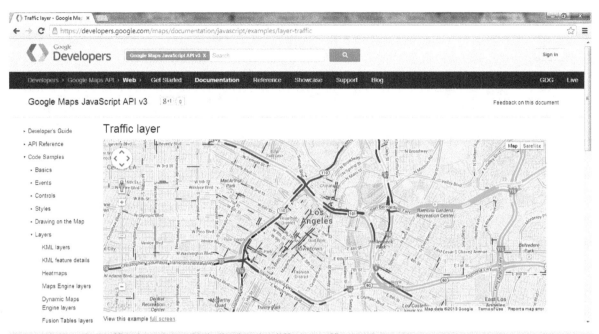

Figure 6. Google Traffic: the color codes indicating the different traffic speeds.

4.1.6. Volunteering Geographic Information

A capability has been provided for users sensing an earthquake to report the event. A form is provided for users to report their name, zip code, telephone, description, latitude, and longitude. Coming up with the results for the Mercalli's intensity levels in case of a severe earthquake with damage reported from survivors, reports from media and first responders. The VGI information input to the system could be used to compliment the Mercalli's intensity levels data and serve as an early first approximation to the Mercalli's scale. The geographic patterns of damage observed from the VGI system can be analyzed and could be used for planning and mitigation in the future.

4.2. Tools

The development environment was set up using the following software:
- Apache Tomcat7, an open source Web server was used to offer a simple, fast and stable performance. It is the most commonly used Web server software and can support programming in Java and a host of other programming languages.
- Eclipse IDE (www.eclipse.org) was used as the integrated development environment (IDE). Eclipse has many features for creating and testing the code. It is one of the most commonly used IDE due to its rich java development tools support and allows third party functional integration [42].
- MySQL [43] was used because it's free and a popular choice for Web application.

4.3. System

The application was implemented using browser-side mash up architecture. Several browser-side APIs (Google API for JavaScript, Google Traffic API for JavaScript and Google Directions API for JavaScript) were used. Dojo Toolkit was used to provide a good user experience. The USGS earthquake feeds were from a RESTful Web service and were in GeoJSON Format. The HTML5 Geolocation tool was used to find the users location from the browser. The Geolocation tool is supported on all the latest browsers like Internet Explorer, Firefox, Chrome, Safari and Opera.

The facilities data was stored in the MYSQL database, from where the data were retrieved using Java was used as the server side programing language and converted into JSON. The facilities within a twenty mile buffer can be calculated by first finding the distance between the user's location and the facilities location. In this project only those facilities within a 20 mile buffer were selected.

The application was deployed on the Amazon AWS EC2 (Elastic Cloud Computing) cloud. Amazon AWS allows a one year free account to all users. A Linux Ubuntu 13.0 64 bit instance along with Tomcat 7 server and MySql database instance needs to be installed. The port settings should be configured so as to allow communication with the cloud. The project can be exported from the local machine to the cloud using FileZilla which is a free and cross platform FTP software. The EIM tool deployed using Amazon EC2 [44], can be found at the following URL http://54.200.221.135:8080/DisasterMashup/ (as of October 5, 2014).

5. System Demonstration

Let us consider a scenario where a 6.3 magnitude earthquake has occurred in the Northern Virginia area, with its epicenter in Fairfax. Damage is reported in the Northern Virginia and DC area. This area has a dense urban environment that includes residents, commuters and visitors from other neighboring areas. The Northern Virginia, DC area, is also home and headquarters to many Federal and National institutions. Accordingly, numerous resources are available for natural disaster assistance, which includes top ranking local response teams, and national agencies with a vast network in place in the event of a natural disaster.

People need not look too far and can make use of the EIM tool as long as they have access to the Internet and consent to provide their location to the EIM tool. Since the EIM tool makes use of the Geolocation API in HTML5, the application asks permission from the user whether they would like to share their location details. Once the user approves to share their location, the application will be able to track their browser's location and their position will be displayed on the map with a marker showing their coordinates. By default, every time the map is loaded, the map displays the current earthquakes in the US for the past week with a magnitude higher than 2.5. The EIM tool displays properties about the earthquake such as the location of the focus of the earthquake along with other properties such as the latitude and longitude, its magnitude, and time at which it occurred.

The application also integrates live traffic conditions for the entire US. If after the Fairfax earthquake a user would like to visit loved ones, she/he could take a look at the EIM tool which provides the current traffic conditions. It would give the user an idea of which roads are currently congested where the traffic is moving very slowly, and they could avoid these roads.

The user can select the VGI option and get a feel of the current situation in Fairfax area provided other users have been reporting events using this tool. The user can also be a citizen reporter and describe the situation and damage in their area. The user of this application will have to provide their Name, Zip code, Telephone number, Description of the event and their latitude and longitude (**Figure 7**). Once the user's entry is approved by the

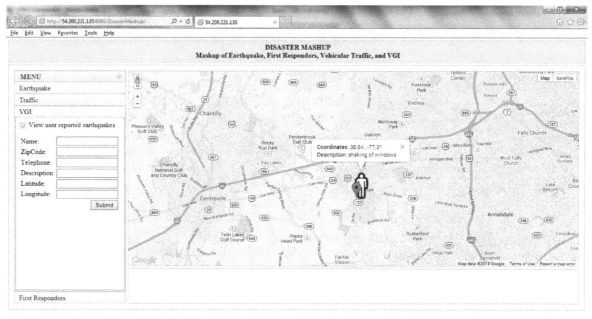

Figure 7. VGI report of events and a description of the scenario.

administrator, the user's entry will be displayed. To protect the identity of the user, only their location and description of event entries are displayed.

The following demonstrates the usage of the system with a few scenarios: 1) finding a hospital close by, 2) finding fire station for emergency, 3) finding police station for public safety, 4) getting help for maintenance.

5.1. Finding a Nearby Hospital for Medical Emergency

Sample challenge: There have been numerous casualties, remaining unattended in the community where the user lives and need to be transported to the nearest hospital.

Analysis: The tool can be used to find out where the medical care facilities in the area are (**Figure 8**). The user can see the medical facilities within a twenty mile radius and choose any one of them. When a user clicks the information window of the medical facilities, contact information is displayed.

Solution: The user could use the telephone number to contact the medical care facility or they could drive directly to the medical care center. The application provides a routing tool to all the medical care facilities within the twenty mile buffer while giving directions and taking into consideration the live traffic conditions.

5.2. Finding a Fire Station

Sample challenge: Houses have collapsed and trees have been uprooted in the vicinity of the user. The roof of the neighboring house has collapsed. Since there is no sign of the neighbors they could be trapped in the debris.

Analysis: The user can use the tool to find local fire stations (**Figure 9**). The user could also see ones within a broader buffer zone, and choose any one of them.

Solution: On clicking the desired fire station, the user is able to obtain the address and contact information. They could call the selected fire station or they could even drive to the fire station and file a complaint. The application provides a routing tool to all the fire station facilities within a twenty mile buffer while taking into consideration live traffic conditions.

5.3. Finding Police Station for Public Safety

Sample challenge: There has also been a lot of widespread looting in the aftermath of the event. The roof of the nearby ABC supermarket has collapsed. Taking advantage of the situation, people are helping themselves to the goods.

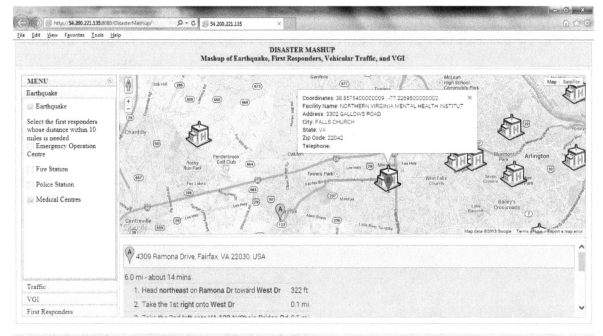

Figure 8. Routing from current location to medical care facilities.

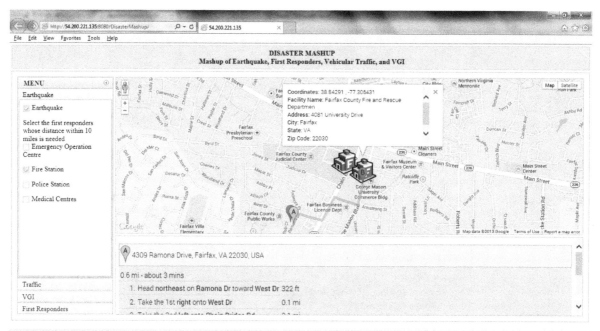

Figure 9. Routing from current location to fire station.

Analysis: The user can use the tool to find out the police stations in the area (**Figure 10**). They could locate the ones within a twenty mile buffer and choose any one of them which they feel would be the best one.

Solution: On clicking the police station the user could obtain the address and contact information. They could either call or drive to the police station to file a complaint. The EIM tool provides a routing tool to all the police station facilities within the twenty mile buffer while taking into consideration the live traffic condition.

5.4. Getting Help for Maintenance

Sample challenge: It has been days since the earthquake occurred. There have been uprooted trees hanging dangerously close to electrical wires *wires*. There has been a power outage power for the past week. The local fire station has not been responding to repeated calls and there has not been help from the county services either.

Analysis: The user can use the EIM tool to find out the emergency operation centers in the area (**Figure 11**). They could also see the ones within the twenty mile buffer and choose any one of them which they feel would be the best one or the closest *closest*.

Solution: On clicking the information window of the emergency operation center they could obtain the address and contact information. They could either call or drive to the emergency operation center to file a complaint and obtain emergency food supplies for themselves.

6. Conclusions and Discussion

This paper reports a Web-based situational awareness earthquake information mashup tool. A prototype was developed to demonstrate the potential efficiency of the system to produce an automated visualization and routing tool for the general public. The system includes a Web based situation awareness GIS that could monitor and map earthquakes and location of emergency facilities, provide live traffic conditions and route path to the emergency facilities. The project shows how a variety of Web technologies and services can be integrated into a scalable visualization model to provide better situation awareness to the user.

Some individuals do not have a good spatial sense and need directions when driving or moving. Even for getting from point A to point B, we require a GPS to guide us there. Similar systems have been used in the past even to provide an accessibility tool to the blind and the visually impaired [45]. A mashup tool can offer a platform which merges data from different sources and has different characteristics. By creating a layer stack of information the tool can create a visualization platform for discerning patterns that exist in the data [46]. It provides the user better situation awareness and the user is able to understand how events around him will impact or

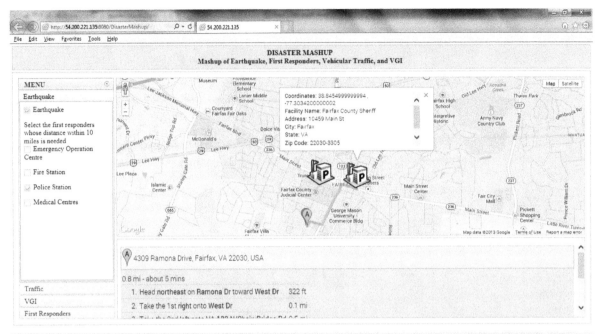

Figure 10. Routing from current location to police station.

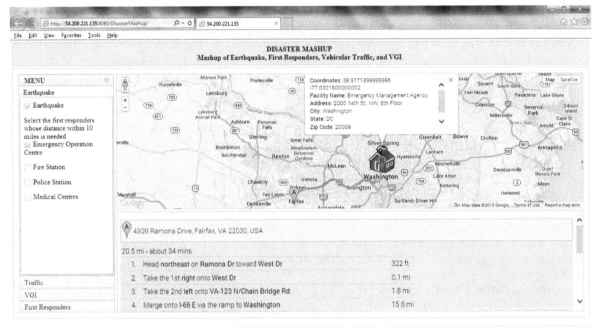

Figure 11. Routing from current location to emergency operation center.

play a role in the near future.

The EIM tool could provide buffer analysis for finding all emergency facilities within a twenty mile buffer and provide a routing tool from the user's current position to these facilities. The EIM tool demonstrated how the Google Maps API, Google Traffic API and Google Directions API for JavaScript could be leveraged through mashup. The Google Maps API for JavaScript provided programmable access to live traffic and directions to facilitate routing to emergency facilities from the users location in live traffic conditions which other map API services like Bing Maps and Open Layers is not able to provide.

During the government shutdown from October 1st through 16th 2013, there was no access to the live earthquake feeds from USGS. At that time other sources for the earthquake feeds were considered. However, we decided to continue using USGS earthquake feeds since the USGS operates more sensors and have better coverage of local areas of interest.

Data used in this application was from Governmental sources like USGS and FEMA, which provides authoritative content; data from Google which provides semi authoritative commercial content; and VGI input, which is asserted content. A quality assessment of the various data sources should be performed in the future research.

The EIM tool demonstrated how VGI can be used to report destruction and damage from a natural disaster. The time when an entry is made should be incorporated so that user can determine what information provided by the system is temporally relevant. Checking the quality of VGI input can be difficult, especially when there is a flood of VGI input messages after an earthquake, but quality assessment is important and there are many methods that can be used, as discussed by Many researchers [47]-[50]. Checking for the accuracy, error and quality of VGI input is therefore a good topic to consider for future research.

Cartographic research should be conducted to take care of overlapping markers in case of earthquakes, and best practices from other cartographic interfaces will be considered, and where appropriate, adopted.

After an earthquake, there could be a flood of users accessing the application. Future research should be undertaken so that the application is able to support a large number of users and still give good performance [51]. Scalability should be considered and techniques such as load balancing, compression, caching and indexing techniques [7] for the database could be investigated to improve the performance utilizing the latest geocomputing technologies, such as spatial cloud computing [52].

The application could be designed for a mobile interface. The HTML5 Geolocation tool could be used to more effect by making use of the other functions which would enable the platform to behave like a GPS and return the updated position of the user as they move or point of interest input by the end user.

Natural disasters such as wildfires, tornadoes could also be incorporated to make it a disaster information mashup system. Other near real time data which also play an important role in the situation like weather and demographic information could be incorporated to show new relationships and perform various analyses like population affected.

Acknowledgements

Zhenlong Li, Kai Liu and Zhipeng Gui helped with the development and deployment of the applications onto cloud servers. Manzhu Yu helped format the paper.

References

[1]   Cutter, S.L. (2003) GI Science, Disasters, and Emergency Management. *Transactions in GIS*, **7**, 439-446.
      http://dx.doi.org/10.1111/1467-9671.00157

[2]   Nolen-Hoeksema, S. and Morrow, J. (1991) A Prospective Study of Depression and Posttraumatic Stress Symptoms after a Natural Disaster: The 1989 Loma Prieta Earthquake. *Journal of Personality and Social Psychology*, **61**, 115-121.
      http://dx.doi.org/10.1037/0022-3514.61.1.115

[3]   Hafner, K. and Lyon, M. (1999) Where Wizards Stay up Late: The Origins of The Internet. Simon and Schuster.

[4]   Crooks, A., Croitoru, A., Stefanidis, A. and Radzikowski, J. (2013) #Earthquake: Twitter as a Distributed Sensor System. *Transactions in GIS*, **17**, 124-147. http://dx.doi.org/10.1111/j.1467-9671.2012.01359.x

[5]   Live Earthquake (2013) Live Earthquake Mashup. http://www.oe-files.de/gmaps/eqmashup.html

[6]   Stefanidis, A., Crooks, A. and Radzikowski, J. (2013) Harvesting Ambient Geospatial Information from Social Media Feeds. *GeoJournal*, **78**, 319-338. http://dx.doi.org/10.1007/s10708-011-9438-2

[7]   Zook, M., Graham, M., Shelton, T. and Gorman, S. (2010) Volunteered Geographic Information and Crowdsourcing Disaster Relief: A Case Study of the Haitian Earthquake. *World Medical & Health Policy*, **2**, 7-33.
      http://dx.doi.org/10.2202/1948-4682.1069

[8]   Vieweg, S., Hughes, A.L., Starbird, K. and Palen, L. (2010) Microblogging during Two Natural Hazards Events: What Twitter May Contribute to Situational Awareness. *Proceedings of the SIGCHI Conference on Human Factors in Computing Systems, CHI'10*, Atlanta, 10-15 April 2010, 1079-1088. http://doi.acm.org/10.1145/1753326.1753486

[9]   Goodchild, M.F. and Glennon, J.A. (2010) Crowdsourcing Geographic Information for Disaster Response: A Research Frontier. *International Journal of Digital Earth*, **3**, 231-241. http://dx.doi.org/10.1080/17538941003759255

[10] Slocum, T.A. (2009) Thematic Cartography and Geovisualization. Pearson Prentice Hall, Upper Saddle River.

[11] FEMA, H.-M. (2013) Hazus | FEMA.gov. http://www.fema.gov/hazus

[12] Petersen, J.F., Sack, D.I. and Gabler, R.E. (2011) Physical Geography. 10th Edition, Cengage Learning.

[13] Wald, D.J., Quitoriano, V., Dengler, L.A. and Dewey, J.W. (1999) Utilization of the Internet for Rapid Community Intensity Maps. *Seismological Research Letters*, **70**, 680-697. http://dx.doi.org/10.1785/gssrl.70.6.680

[14] Longley, P. (2005) Geographic Information Systems and Science. John Wiley & Sons, Hoboken.

[15] Johnson, R. (2011) Russ Johnson Talks Esri and All Things GIS.
http://www.emergencymgmt.com/disaster/Russ-Johnson-Talks-Esri-GIS.html?page=2&

[16] O'Reilly, T. (2005) What Is Web 2.0: O'Reilly Media.
http://oreilly.com/pub/a/Web2/archive/what-is-Web-20.html?page=1

[17] Kraak, M.J. (2007) Geovisualization and Visual Analytics. *Cartographica: The International Journal for Geographic Information and Geovisualization*, **42**, 115-116. http://dx.doi.org/10.3138/carto.42.2.115

[18] Sun, M., Li, J., Yang, C., Schmidt, G.A., Bambacus, M., Cahalan, R., Huang, Q., Xu, C., Noble, E.U. and Li, Z. (2012) A Web-Based Geovisual Analytical System for Climate Studies. *Future Internet*, **4**, 1069-1085.
http://dx.doi.org/10.3390/fi4041069

[19] Fu, P. and Sun, J. (2011) WebGIS: Principles and Applications. ESRI Press, Redlands.

[20] Goodchild, M.F. (2007) Citizens as Sensors: The World of Volunteered Geography. *GeoJournal*, **69**, 211-221.
http://dx.doi.org/10.1007/s10708-007-9111-y

[21] Yang, C., Fu, P., Goodchild, M.F. and Xu, C. (2014) Integrating GIScience Applications through Mashup. CyberGIS, in press.

[22] Batty, M., Hudson-Smith, A., Milton, R. and Crooks, A. (2010) Map Mashups, Web 2.0 and the GIS Revolution. *Annals of GIS*, **16**, 1-13. http://dx.doi.org/10.1080/19475681003700831

[23] Butler, D. (2006) Mashups Mix Data into Global Service. *Nature*, **439**, 6-7. http://dx.doi.org/10.1038/439006a

[24] Liu, S.B., Palen, L., Sutton, J., Hughes, A.L. and Vieweg, S. (2008) In Search of the Bigger Picture: The Emergent Role of On-Line Photo Sharing in Times of Disaster. *Proceedings of the Information Systems for Crisis Response and Management Conference (ISCRAM)*. http://works.bepress.com/vieweg/11

[25] Maceachren, A.M. and Brewer, I. (2004) Developing a Conceptual Framework for Visually-Enabled Geocollaboration. *International Journal of Geographical Information Science*, **18**, 1-34.
http://dx.doi.org/10.1080/13658810310001596094

[26] MacEachren, A.M., Gahegan, M., Pike, W., Brewer, I., Cai, G., Lengerich, E. and Hardistry, F. (2004) Geovisualization for Knowledge Construction and Decision Support. *IEEE Computer Graphics and Applications*, **24**, 13-17.
http://dx.doi.org/10.1109/MCG.2004.1255801

[27] Paulson, L.D. (2005) Building Rich Web Applications with Ajax. *Computer*, **38**, 14-17.
http://dx.doi.org/10.1109/MC.2005.330

[28] Gui, Z., Yang, C., Xia, J., Li, J., Rezgui, A., Sun, M., Xu, Y. and Fay, D. (2013) A Visualization-Enhanced Graphical User Interface for Geospatial Resource Discovery. *Annals of GIS*, **19**, 109-121.
http://dx.doi.org/10.1080/19475683.2013.782467

[29] USGS Earthquake, Ha. P. (2013) 1 Day, Magnitude 2.5+ Worldwide. http://earthquake.usgs.gov/earthquakes/map/

[30] USGS, Live Feeds (2013) Real-Time Feeds & Notifications. http://earthquake.usgs.gov/earthquakes/feed/v1.0/

[31] RSOE, E. (2013) RSOE EDIS: Emergency and Disaster Information Service.
http://hisz.rsoe.hu/alertmap/index2.php?area=usa

[32] IRIS Earthquake B. (2013) IRIS Earthquake Browser. http://www.iris.edu/ieb/

[33] REV, E.V. (2013) REV: Earthquake View. http://rev.seis.sc.edu/earthquakes.html?view=contig_us

[34] ESRI, P. I. M. (2013) Earthquakes. http://tmappsevents.esri.com/Website/earthquake-responsive/

[35] Google (2013) Too Many Markers! Google Maps API—Google Developers.
https://developers.google.com/maps/articles/toomanymarkers

[36] Dojo (2013) Unbeatable JavaScript Tools—The Dojo Toolkit. http://dojotoolkit.org/

[37] Putz, S. (1994) Interactive Information Services Using World-Wide Web Hypertext. *Computer Networks and ISDN Systems*, **27**, 273-280. http://dx.doi.org/10.1016/0169-7552(94)90141-4

[38] Yang, C., Wong, D.W., Yang, R., Kafatos, M. and Li, Q. (2005) Performance-Improving Techniques in Web-Based GIS. *International Journal of Geographical Information Science*, **19**, 319-342.
http://dx.doi.org/10.1080/13658810412331280202

[39] Google Maps (2013) Getting Started: Google Maps JavaScript API v3—Google Developers.
https://developers.google.com/maps/documentation/javascript/tutorial

[40] Google Traffic API (2013) Traffic Layer—Google Maps JavaScript API v3—Google Developers. https://developers.google.com/maps/documentation/javascript/examples/layer-traffic

[41] Google Directions (2013) The Google Directions API: Google Maps API Web Services—Google Developers. https://developers.google.com/maps/documentation/directions/

[42] Murphy, G.C., Kersten, M. and Findlater, L. (2006) How Are Java Software Developers Using the Elipse IDE? *IEEE Software*, **23**, 76-83. http://dx.doi.org/10.1109/MS.2006.105

[43] MYSQL (2013) MySQL : The World's Most Popular Open Source Database. http://www.mysql.com/

[44] Amazon, EC2. (2013) AWS | Amazon Elastic Compute Cloud (EC2)—Scalable Cloud Servers. http://aws.amazon.com/ec2/

[45] Rice, M.T., Aburizaiza, A.O., Jacobson, R.D., Shore, B.M. and Paez, F.I. (2012) Supporting Accessibility for Blind and Vision-Impaired People with a Localized Gazetteer and Open Source Geotechnology. *Transactions in GIS*, **16**, 177-190. http://dx.doi.org/10.1111/j.1467-9671.2012.01318.x

[46] MacEachren, A.M. (2004) How Maps Work: Representation, Visualization, and Design. Guilford Press, New York.

[47] Goodchild, M. and Li, L. (2012) Assuring the Quality of Volunteered Geographic Information. *Spatial Statistics*, **1**, 110-120. http://dx.doi.org/10.1016/j.spasta.2012.03.002

[48] Girres, J. and Touya, G. (2010) Quality Assessment of the French OpenStreetMap Dataset. *Transactions in GIS*, **14**, 435-459. http://dx.doi.org/10.1111/j.1467-9671.2010.01203.x

[49] Haklay, M. (2010) How Good Is Volunteered Geographical Information? A Comparative Study of OpenStreetMap and Ordnance Survey Datasets. *Environment and Planning B: Planning & Design*, **37**, 682-703. http://dx.doi.org/10.1068/b35097

[50] Rice, M., Paez, F., Mulhollen, A., Shore, B. and Caldwell, D. (2012) Crowdsourced Geospatial Data: A Report on the Emerging Phenomena of Crowdsourced and User-Generated Geospatial Data. Annual. George Mason University, Fairfax. http://www.dtic.mil/dtic/tr/fulltext/u2/a576607.pdf

[51] Li, Z., Yang, C.P., Wu, H., Li, W. and Miao, L. (2011) An Optimized Framework for Seamlessly Integrating OGC Web Services to Support Geospatial Sciences. *International Journal of Geographical Information Science*, **25**, 595-613. http://dx.doi.org/10.1080/13658816.2010.484811

[52] Yang, C., Xu, Y. and Nebert, D. (2013) Redefining the Possibility of Digital Earth and Geosciences with Spatial Cloud Computing. *International Journal of Digital Earth*, **6**, 297-312. http://dx.doi.org/10.1080/17538947.2013.769783

12

Designing an Online Geospatial System for Forest Resource Management

Peter G. Oduor[1], Michael Armstrong[2], Larry Kotchman[3], Michael Kangas[4], Buddhika Maddurapperuma[5], Kelsey Forward[6], Pubudu Wijeyaratne[7], Xiana Santos[8], Akiko Nakamura[9], Krystal Leidholm[10]

[1]Department of Geosciences, North Dakota State University, Fargo, USA
[2]IT Systems Vulnerability, Sanford Health Fargo, Fargo, USA
[3]North Dakota Forest Service, Bottineau, USA
[4]North Dakota Forest Service, North Dakota State University, Fargo, USA
[5]Environmental & Conservation Science Program, North Dakota State University, Fargo, USA
[6]Wenck Associates North Dakota, Fargo, USA
[7]Department of Computer Science, North Dakota State University, Fargo, USA
[8]Department of Forestry, Mississippi State University, Starkville, USA
[9]East View Geospatial, Minneapolis, USA
[10]California Department of Fish & Game, Washburn, USA
Email: Peter.Oduor@ndsu.edu, Mike.Armstrong@sanfordhealth.org, Larry.Kotchman@ndsu.edu, Michael.Kangas@ndsu.edu, b.madurapperuma@my.ndsu.edu, kforward@wenck.com, pubudu.wijeyaratne@ndsu.edu, xsantos@CFR.MsState.edu, Akiko.Nakamura@eastview.com, kleidholm@hotmail.com

Abstract

Geographic and Geospatial information systems (GISs) have especially benefited from increased development of their inherent capabilities and improved deployment. These systems offer a wide range of services, for example, user-friendly forms that interact with the geospatial components for locational information and geographic extents. An online distributed platform was designed for forest resource management with map elements residing on a GIS platform. This system is accessible on non-authenticated browsers optimized for desktops; whereas the online resource management forms are also accessible on mobile platforms. The system was primarily designed to aid foresters in implementing resource management plans or track threats to forest resource. Baseline data from the system can be easily visualized and mapped. Other data from the system-can provide input for stochastic analyses especially with respect to forest resource management.

Keywords

ArcGIS Server, Wildland Fire, Forest Stewardship, North Dakota

1. Introduction

Decision Support Systems (DSSs) are fundamental in addressing complexity of making coherent, integrated, and interdependent resource management decisions. This is due to their inherent nature of ability to cohesively formulate those parameters or pertinent information that otherwise cannot be processed effectively by human heuristic processes. Decisions formulated from DSSs must be defensible by stakeholders (e.g. [1]), factor in multi-scalability and temporal issues, factor in other relevant considerations, aid in resolving potential conflicts amongst other factors. Interactive computer-based systems have been adopted to help decision makers utilize data and models to solve unstructured problems or decision support systems [2] [3]. DSSs have evolved to encompass multi-component systems that include various combinations of simulation modeling, optimization techniques, heuristics and artificial intelligence techniques, geographic information systems (GIS), associated databases for calibration and execution, and user interface components [4]. Each of these six components may to some degree individually satisfy Sprague and Carlson's [3] generic DSS definition.

An Adaptive Decision support systems (ADSS) may be interpreted to include any system that is capable of self-teaching, which is accomplished by integrating unsupervised inductive learning methods (e.g. [5]-[7]). ADSSs reduce effectively the need for implementing complex spatial analytical capabilities on an ArcGIS server platform by generating the best result to a problem by refining an initial solution. This can be done by essentially incorporating the results of the spatial analyses as a layer with identifiable features. Adaptability of such a system may arise from GUI designs (dialog subsystem) with pertinent factors accounted for (e.g. [8]-[11]), degree of interactivity of data (database subsystem), auxiliary information (knowledge subsystem), spatial analyses (problem processing subsystem), statistically derived data (model base subsystem) and lastly expert analyses (decision-based subsystem).

The primary aim of this study was to integrate an out-of-the-box ArcGIS Server system with hallmarks characteristic of an ADSS essentially to: 1) provide a secure gateway for the North Dakota SAP derived data layers for resource management by depicting lands rich in natural resources, vulnerable to threat or both. 2) Serve as a city, county, state and federal reporting mechanism for NDFS and affiliated partners' forestry management accomplishments; a resource locale identifier; a tracking and monitoring geospatial interface; and a public resource to monitor or track threats or vulnerabilities including invasive species, catastrophic wildfire, and climate change effects on forestry and forest conversion. 3) Provide baseline data on the forest resources of North Dakota to model potential forest resource threats and offer management opportunities identified through the state forest resource assessment and/or vulnerabilities identified using the State and Private Forestry Redesign national assessment tool. 4) Determine and identify tracts that were not included in the original spatial analysis project. 5) Provide a concise central repository and inventory of forestry programs identifiable by searchable attributes such as city, county or associated wild land-urban interface. The specific objective was to design a widely distributed online portal for forest resource management.

2. Overview of North Dakota

North Dakota was formerly the northern portion of Dakota Territory, located in the Midwestern region of the United States, became a US state in 1889. It borders Minnesota to the east, South Dakota to the south, Montana to the west and Canadian provinces of Manitoba and Saskatchewan to the north. It spans a latitudinal range of 45.93° - 49° and extends westwards from 96.55° to 104.05° longitude. Its areal coverage makes it the 19th most extensive US state and comprises of 53 counties (**Figure 1**). North Dakota state capitol is located in Bismarck on the banks of the Missouri River just downstream from Lake Sakakawea. Lake Sakakawea, a large man-made lake, is behind Garrison Dam. The largest city is Fargo on the banks of the Red River. North Dakota was considered part of the Great American Desert. A precipitation gradient exists from east to west. The eastern regions generally receive more precipitation. The area in the past was resplendent with devastating prairie fires making

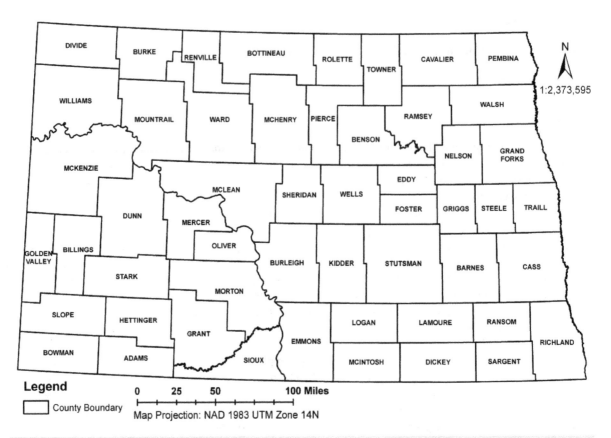

Figure 1. An overview map of North Dakota, USA.

establishment of arboreal ecosystems extremely difficult. The western half of the state consists of the hilly Great Plains, and the northern part of the Badlands to the west of the Missouri River [12] [13]. The state's high point, White Butte at 3506 feet (1069 m), and Theodore Roosevelt National Park [12] [13] are located in the Badlands. North Dakota is abundant in fossil fuels, for example, natural gas, crude oil, lignite coal predominantly in the western part of the state. Fossil fuels form primary economic activities in the western part of the state whereas the eastern part of the state has a thriving agricultural sector industry. Natural trees in North Dakota include riparian forests around perennial streams, around Killdeer and Turtle Mountains, and in significant plantings, for example, managed forests and in other areas such as shelterbelts. North Dakota forests are comprised of four major types: elm-ash-cottonwood, aspen, oak and ponderosa pine [12]. The North Dakota Forest Service (NDFS) was established in 1906 to practice sound land stewardship to enhance and preserve forests, grassland, and wetland ecosystems found within the state boundaries [12]. By 1954, the total acreage for protection plantings was 89,000 acres (360,170,221.59 m^2), earning North Dakota the distinction of having more protection plantings than any other state in the United States. To date, the natural woodlands of North Dakota covered about 824,000 acres of forested land that includes shelterbelts.

3. Methodology

3.1. Phase I: System Structure

The system design entails three user levels accessible to: 1) local users and administrators, 2) registered foresters, and 3) rural fire departments, researchers and the general public (**Figure 2**). Administrators are able to view log files, update databases, update the geospatial database elements, register new users and perform system maintenance tasks. The design schema has two main databases accessible to 1) foresters (private database), and 2) public (public database) with pertinent security and protection mechanisms instituted. The public database is accessible through non-authenticated browsers (**Figure 3**).

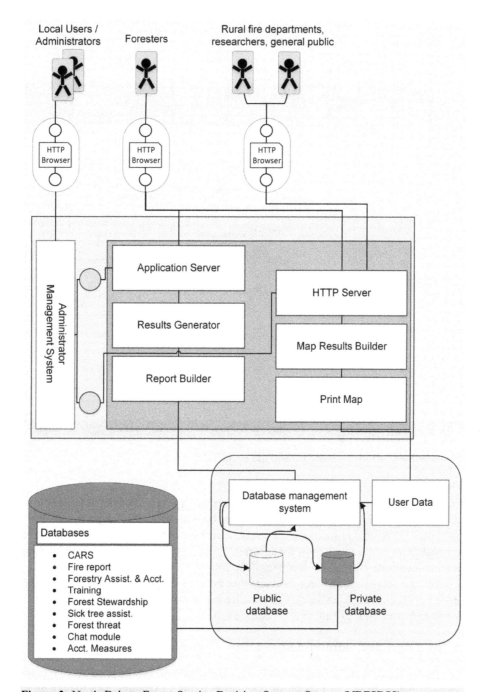

Figure 2. North Dakota Forest Service Decision Support System (NDFSDSS) system configuration.

3.2. Phase II: Management System

Several databases were linked to web forms to capture resource management data. The forms were designed to, capture user information, represent underlying features integral to the database, for example, associated FIPS codes, store the dialog or knowledge base, and provide on demand pertinent information from multiple sources for forestry resource management. For example, for Forest Stewardship plan, a dynamic interaction between the user and the system enables amongst other alternatives, ability to determine if a proposed plan would be in a prioritized area. We also implemented minimum data storage and applied security measures in retrieval of non-sensitive data. Most of the forms designed are available to the general public, while the databases can only

Table of Contents   Map Navigation Tools & Buttons

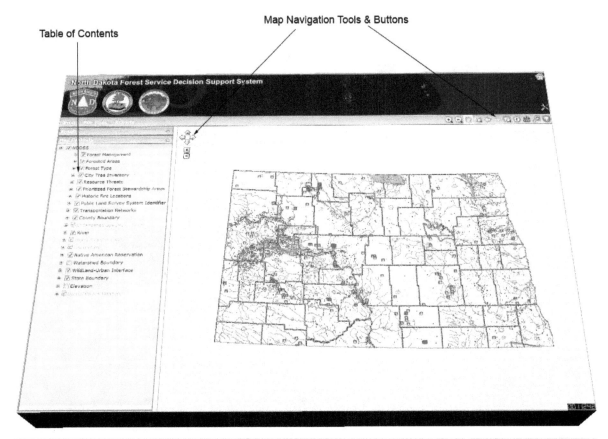

Figure 3. Opening page of NDFSDSS (http://ndfsdss.dakotacollege.edu/).

be accessed by the North Dakota Forest Service personnel.

Types of forms and databases designed include: 1) Community Accomplishments Reporting System For Urban And Community Forestry Program (form), 2) North Dakota Rural Fire Department Wild land Fire Report (form and database). 3) Redesign-Innovation in State and Private Forestry form, 4) Technical Forestry Assistance and Accountability Measures Report for Information and Education (form and database), 5) Training Program/Presentation Template (form and database). The following forms require login: 1) Accountability Measures Report (form and database). 2) Forest Stewardship Program and Rural Forestry Assistance (form). 3) Sick Tree Assistance Form (form and database). 4) Forest Resource Management Plan (template). 5) Report a Forest Threat (template and database). 6) Talk to a Forester (chat).

3.3. Community Accomplishments Reporting

The CARS (Community Accomplishments Reporting System for Urban & Community Forestry Program) allows for measured outcomes on: 1) Percent of population living in communities managing programs to plant, protect and maintain their urban and community trees and forests. 2) Percent of population living in communities developing programs and/or activities to plant, protect and maintain their urban and community trees and forests. The designed form allows for multi-year tracking of these percentages to gauge community participation and success of community based programs for various stakeholders. The form accepts all generic CARS related Microsoft Excel® comma delimited files. Various outputs can also be tracked for each fiscal year. These include: 1) Number of people living in communities provided educational, technical and/or financial assistance. 2) Number of people living in communities that are developing programs/activities for their urban and community trees and forests. 3) Number of people living in communities managing their urban and community trees and forests. 4) Number of communities with active urban & community tree and forest management plans developed from professionally-based resource assessments/inventories. 5) Number of communities that employ or retain through written agreement the services of professional forestry staff who have at least one of these credentials: a) degree in forestry or related field and b) ISA certified arborist or equivalent professional certification. 6) Number of communities that have adopted and can present documentation of local/statewide ordinances or policies that focus on planting, protecting, and maintaining their urban and community trees and forests. 7) Number of communities with local advocacy/advisory organizations, such as, active tree boards, commissions, or non-profit organizations that are formalized or chartered to advise and/or advocate for the planting, protection, and maintenance of urban and community trees and forests. 8) Number of hours of volunteer service logged. (An agency-wide consistent methodology to be developed to track volunteer hours). 9) State offered community grant program in current fiscal year. 10) Number of communities receiving financial assistance awarded during the Federal FY 2010 through a state managed community grant program. 11) Amount of Federal (USFS) funding to States. From the database critical needs may be addressed especially for: a) communities that have the potential to develop management programs for their trees and forests with assistance from UCF technical, financial and/or educational program services, and b) communities that currently are not managing, or developing programs to manage, their urban and community trees and forests. Finally, an estimate of federal (USFS) dollar cost or expenditure per capita in community assisted can be tracked.

3.4. Fire Report

This form was designed to collate forest fire information for wildland fire occurrences within North Dakota. Pertinent data collected include: 1) Fire discovery and containment dates, 2) fire size (acreage), 3) locational information (latitude and longitude and land ownership, 4) cause of fire, 5) vegetation burned, and 6) structures lost or threatened. This information will be critical in modeling fire disturbance and spread, for example, by defining a non-parametric separation index (SI) to potentially determine which cover types are prone to fire disturbances. The cover types listed on the form include, grass, cropland, pine forest, hardwood forest, brush and any other category. North Dakota has a climatic gradient from the drier West to well-watered Eastern parts. The SI can be calculated from [14]:

$$\mathrm{SI}_{i,j} = 1 - \frac{A_{i,j}}{\mathrm{Min}\left(A_i, A_j\right)}, \tag{1}$$

where $\mathrm{SI}_{i,j}$ is separation index between cover types i and j $\left(0 \leq \mathrm{SI}_{i,j} \leq 1\right)$, $A_{i,j}$ is the overlap area between cover types i and j, A_i or A_j is area for cover type i or j, and Min represents the minimum function (smaller number between A_i and A_j). Data from the GIS application and fire data can also be used to model spatio-temporal variability in fire return intervals using Stambaugh and Guyette method [15]. Fire return intervals can be estimated from empirically determining a Mean Fire Interval (MFI) from [15]:

$$\mathrm{MFI} = f\left(\mathrm{TRI}, \mathrm{POP}, \mathrm{RD}\right), \tag{2}$$

where TRI denotes the topographic roughness index, POP signifies the natural log of human population density, and RD is river distance. Flame height can be modeled from the following equation [16]:

$$\frac{H_f}{H_{f0}} = \left[1 + \frac{4U_0^2}{gH_{f0}}\right]^{-1/2}, \tag{3}$$

where U_0 is wind speed at a given height (m/s), H_{f0} is flame height (m) in the absence of wind, and g is gravity (m/s^2). The tangent of the flame tilt angle is proportional to $\sqrt{\left(U_0^2/gH_{f0}\right)}$.

3.5. Forestry Assistance and Accountability

North Dakota Forest Service through its outreach programs extends educational components through several programs. These include, Natural Resource Conservation Education, Envirothon, Arbor Day, and Smokey Bear Poster Contests. The extension of this program is facilitated through numerous avenues including the Teacher Learning Centers across the state, Project Learning Tree workshops, and via other K-12 outreach programs. Some of these programs are funded under the auspices of the North Dakota Environmental Education Strategic State Plan. Organizers of these events can now report through the Forestry Assistance & Accountability form. This form is also dynamically-linked to the Accountability Measures form (**Figure 4**). This information can

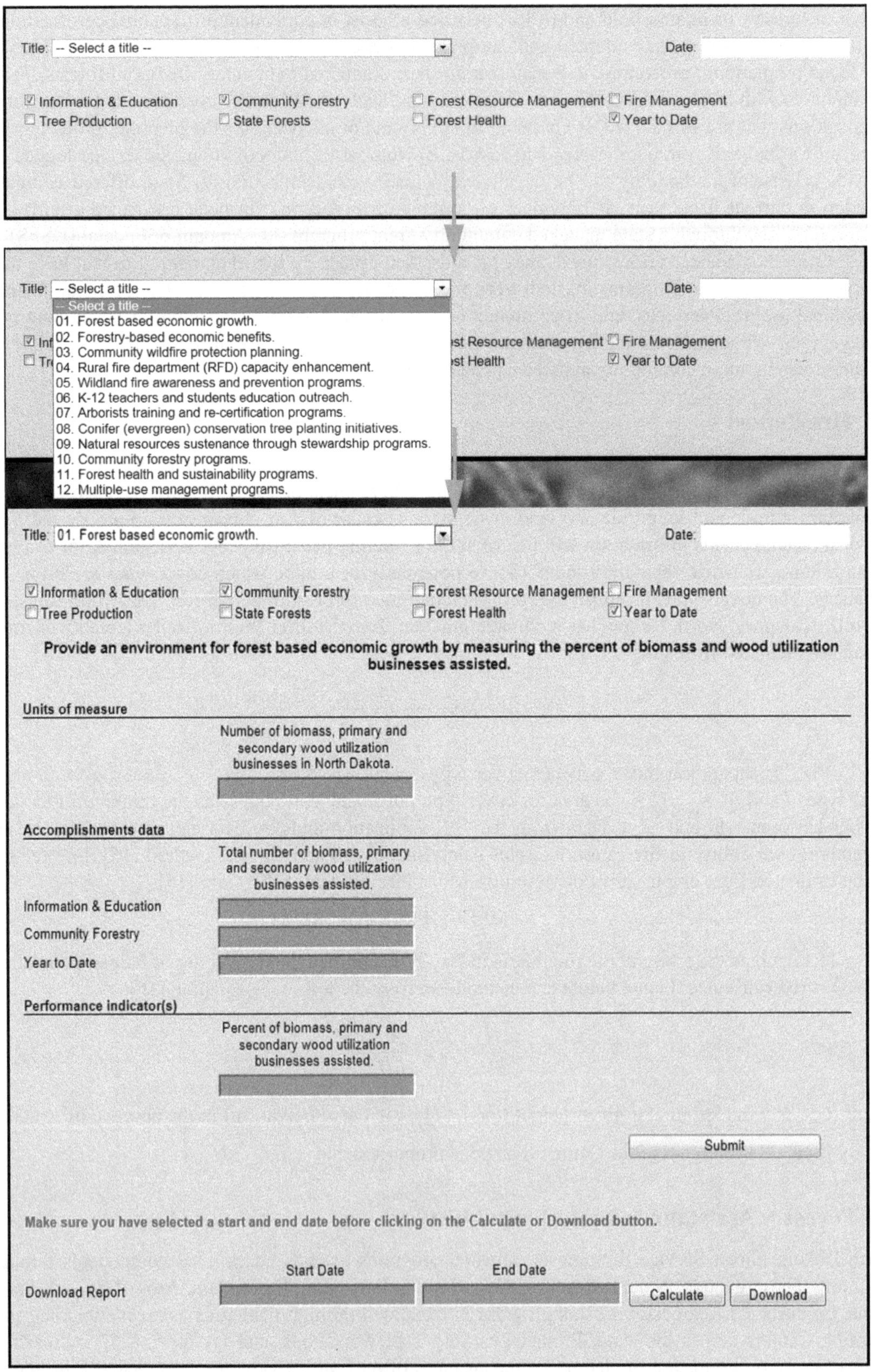

Figure 4. Multi-component Accountability Measures Report form designed to integrate several levels of data in a master RDBMS.

be used to track educational programs across the state.

3.6. Training

The training form populates a database with available training programs and cooperating agencies. Listed training includes 1) Fire department training, 2) Insect and disease training, 3) Landowner education and training, and any other agency-based training program. Cooperating agencies that may be involved include North Dakota Department of Agriculture (NDDA), ND Game & Fish, NDSU Extension, Animal and Plant Health Inspection Service (APHIS), Tribal Organizations amongst other potential agencies.

3.7. Forest Stewardship Program

Original guidelines were meant to delineate potential stewardship tracts within states; provide tools necessary for the North Dakota Forest Service to effectively and efficiently address critical forest resource issues at state, regional, and community scales; and provide forest resource managers with unbiased means to address problems, opportunities and objectives associated with intermingled federal, city, state and private land ownership patterns within North Dakota. The form (**Figure 5**) designed for forest stewardship and rural forestry assistance help track number of landowners that participated in landowner assistance or education-based programs. The value for education-based programs comes from Section 3.6. The form also calculates the acreage for 1) new and/or revised Forest Stewardship Management plans, 2) new and/or revised Forest Stewardship Management Plans that are in prioritized areas. These areas include high priority areas, medium priority, and low priority areas (e.g.

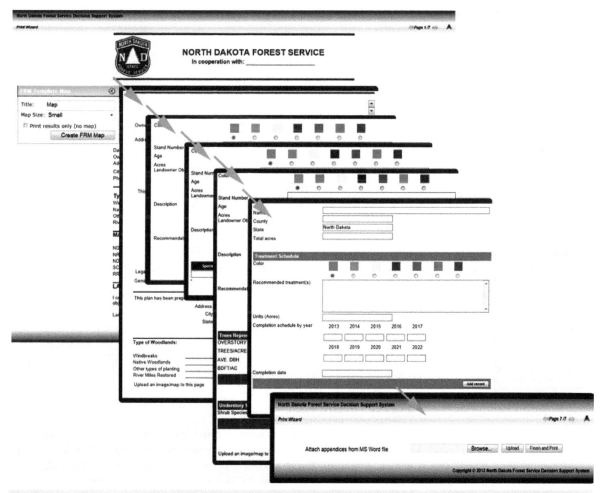

Figure 5. Forest resource management and stewardship template. The designed template provides map interactivity with locations, resource threats/potential and associated base maps.

[10]). The number of plans for any fiscal year can be queried within the database. Other critical areas that can be easily queried include: a) Base Non-Industrial Private Forests (NIPF) acres in important forest resource areas, b) acres covered by current Forest Stewardship plans, c) acres in important forest resource areas covered by current Forest Stewardship plans, d) total number of acres in important forest resource areas being managed sustainably, as defined by a current Forest Stewardship Plan, e) acres currently under an Environmental Quality Incentives Program (EQIP) Management Plan.

Using Markovian random processes e.g. [13] we successfully utilized the Forest Stewardship Program data to model the transition potentials and areal changes for eastern North Dakota. Our basic paradigm was to define forest transition as a first-order Markov process $\{X_t\}$ with $n = \{0,1,2,...\}$ and $t_n \in T$ for $T = \{0,\infty\}$. If the conditional probability is not affected by earlier states then [13] [17]:

$$P_{X(1|n-1)}\left(X_n,t_n \mid X_1,t_1; X_2,t_2;...; X_{n-1},t_{n-1}\right) = P_{X(1|1)}\left(X_n,t_n \mid X_{n-1},t_{n-1}\right), \tag{4}$$

with $P_{X(1|1)}$ as the transition density. A Markovian process can be adequately determined by two functions, $P_{X(1)}\left(X_1,t_1\right)$ and $P_{X(1|1)}\left(X_2,t_2 \mid X_1,t_1\right)$ such that for $t_1 < t_2 < t_3$ then [17]:

$$
\begin{aligned}
P_{X(3)}\left(X_1,t_1; X_2,t_2; X_3,t_3\right) &= P_{X(2)}\left(X_1,t_1; X_2,t_2\right) P_{X(1|2)}\left(X_3,t_3 \mid X_1,t_1; X_2,t_2\right) \\
&= P_{X(1)}\left(X_1,t_1\right) P_{X(1|1)}\left(X_2,t_2 \mid X_1,t_1\right) P_{X(1|1)}\left(X_3,t_3 \mid X_2,t_2\right)
\end{aligned} \tag{5}
$$

which also holds true for all hierarchy of $P_{X(n)}$. The time-stationary Markov chain can be determined by the Markov transition matrix $p_{ij}(t) = P\{X_t = j \mid X_{t-1} = i\} \in \Re^{i \times j} = p_{ij}$ for $i = j = m$ and $t_n \in T$ such that [13] [18]:

$$P(t) = \begin{bmatrix} p_{1,1}(t) & p_{1,2}(t) & \cdots & p_{1,m}(t) \\ p_{2,1}(t) & p_{2,2}(t) & \cdots & p_{2,m}(t) \\ \vdots & \vdots & \ddots & \vdots \\ p_{m,1}(t) & p_{m,2}(t) & \cdots & p_{m,m}(t) \end{bmatrix}, \tag{6}$$

with $\sum_{j=1}^{m} p_{ij} = 1$ and as such [13] [18]:

$$\hat{p}_{ij}(t) = \frac{n_{ij}(t)}{n_i(t-1)} = \frac{n_{ij}(t)}{\sum_{k=1}^{m} n_{ik}(t)} \tag{7}$$

where $n_i(t-1)$ is the total number of cells transiting from category i during the tth transition period, $n_{ij}(t)$ is the number of cells transiting from category i to j in the tth transition period (Wu $et\ al.$, 2006). From Equation (6), if $\lim_{i \to 0+} P(t) = \delta_{ij}$ where $\delta_{ij} = 1$ for $i = j$ and $\delta_{ij} = 0$ otherwise, then $P(t)$ is continuous at $t = 0$ and $\lim_{t \to 0+} \left[\{P(t) - I\}/t\right] = Q = q_{ij} \in \Re^{m \times m}$ where I is the identity matrix, with $q_{ij}(t) \geq 0 \ \forall \ j \neq i$ and $q_{ii}(t) = -\sum_{j \neq i} q_{ij}(t) \ \forall \ t \geq 0$ [13] [19]. Q is infinitesimal generator that represents the rates of change of the transitions. We can therefore surmise that the transition rate at time y, $q_{ij}(y)$ are solutions to, $dp_{ij}(y)/dy = \sum_k p_{ik}(y) q_{kj}(y)$, system of differential equations.

3.8. Sick Tree Assistance, Forest Threat, Chat

The sick tree assistance form was designed to aid private landowners provide base data for managing any tree that exhibited disease symptoms. An exhaustive listing is provided for common tree species within the state with a chance of inputting cultivars if known. The forest threat module was designed to provide a concise detail of potential threats in the following categories: 1) Invasive plants, 2) insects, 3) diseases, 4) climate, 5) loss of open space, 6) pollution, 7) wild land fires, 8) other invasive species, for example, non-native earthworks, 9) unmanaged recreation or any other unlisted threat. The form also provides for each listed threat an image and text description of each threat on computer mouse hover (**Figure 6**). A chat module was also created primarily for foresters from different geographic locales to be able to dialogue especially on pertinent issues. In this way the chat module effectively provides an extra platform for communication.

3.9. Accountability Measures

The accountability measures report is the most comprehensive management component designed. It coprises of twelve sections and seven areas of accomplishments data that could be used in any combination. The sections include: 1) Forest based economic growth. 2) Forestry-based economic benefits. 3) Community wildfire protection planning. 4) Rural fire department (RFD) capacity enhancement. 5) Wildland fire awareness and prevention programs. 6) K-12 teachers and students education outreach. 7) Arborists training and re-certification programs. 8) Conifer (evergreen) conservation tree planting initiatives. 9) Natural resources sustenance through stewardship programs. 10) Community forestry programs. 11) Forest health and sustainability programs. 12) Multiple-use management programs. The seven areas for accomplishments data include: 1) Information and education, 2) Community forestry, 3) Forest Resource Management, 4) Fire management, 5) Tree production, 6) State forests, and 7) Forest health. For each accomplishment data a performance indicator is automatically calculated based on underlying factors within the seven areas and standardized units of measure. A searchable by date database is also generated.

4. Result Analysis

It is worthwhile to note that most wild land fire occurrences are on riparian or close to riparian forests (**Figure 7**). This indicates that even with relatively higher stand densities, decreased probability of torching and lower canopy heights amongst other factors they still had higher probability of torching than upland forests. **Figure 8(a)** shows average and maximum wind speeds for North Dakota Agricultural Weather Network (NDAWN) stations recorded on the same day that wild land fire was recorded. Thiessen polygons were generated using ArcGIS to determine which GPS locations of wild land fires corresponded to which NDAWN station. It is clear that maximum wind speeds were recorded where the burned areas were larger. At Bottineau the burned area exceeded 400 m^2 although the wind speed was comparatively lower, this can be attributed to the type of less fire tolerant vegetation found in this area which is also resplendent with typical needle-leaves trees. From Equation (3), we modeled H_f/H_{f0} for H_{f0} = 0.2 m, 0.4 m, 0.6 m, 0.8 m, 1.0 m, 1.2 m, 1.4 m and 1.6 m. The rationale of choice for H_{f0} values was to cover a broad range of wildland vegetation, that is, prairie grasses that cover most of North Dakota, intermittent shrub heights and also to allow for an acceptable range of anthropogenically-induced wildland fire heights (**Figure 8(b)**). From **Figure 8(c)**, for the H_{f0} values used, we found that H_f/H_{f0} varies from 4% - 59% ($0.042 \le H_f/H_{f0} \le 0.594$). NDAWN stations that had higher ranges for H_f/H_{f0} included Hazen, Turtle Lake and Watford City, however, there burned areas were smaller in acreage. **Figures 9(a)-(h)** show variation of flame height in cm when wind is considered with respect to latitude and longitude. For the range of H_{f0} values used the observed trend is that flame heights with respect to wind are higher in a band trending NW - SE along the central grasslands region. From **Figure 7**, in 2012, most fire occurrences reported for this region were in Stutsman County. It is also worthwhile that there is no discernible uniform trend for flame heights for each location (trend would mimic same trend that can be generated from **Figure 8(c)**). All graphs were generated using SigmaPlot (Systat Software, San Jose, CA). **Figures 10(a)-(h)** represent polar plots where the distance from the origin of the graph is given by $\sqrt{\left(U_0^2/gH_{f0}\right)}$ and the angle between the positive horizontal axis and the radius vector from the origin is represented by the wind direction as recorded by NDAWN stations. They represent an annual composite of wind direction roses with respect to recorded fire events. It can be clearly seen that for locations where the prevailing directions are westward there are correspondingly higher values of $\sqrt{\left(U_0^2/gH_{f0}\right)}$ and as such larger tangential tilts and equally larger areas affected. At an annualized outlook of wind patterns and several locations it is equally harder to predict incoming and return flow directions. Nevertheless, we can adduce from **Figures 10(a)-(h)** that for $0.20 \text{ m} \le H_{f0} \le 1.60 \text{ m}$ there is a prevailing NW trend for wind direction.

From **Figures 11(a)-(h)**, the following categories of land use and land cover are enumerated: 1-Row crops,

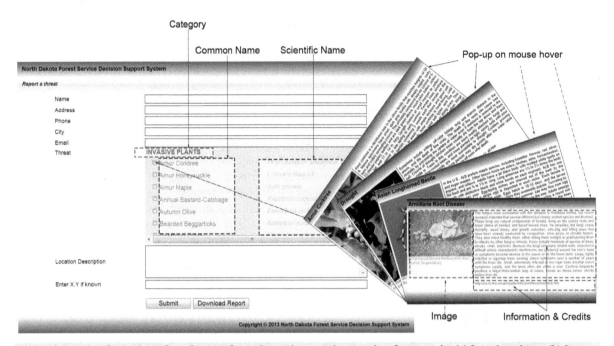

Figure 6. Report a forest threat form features forest threats in several categories, for example; (a) Invasive plants, (b) Insects, (c) Diseases, (d) Climate, (e) Loss of open space, (f) Pollution, (g) Wildland fires, (h) Other invasive species, (i) Unmanaged recreation and an "other" category.

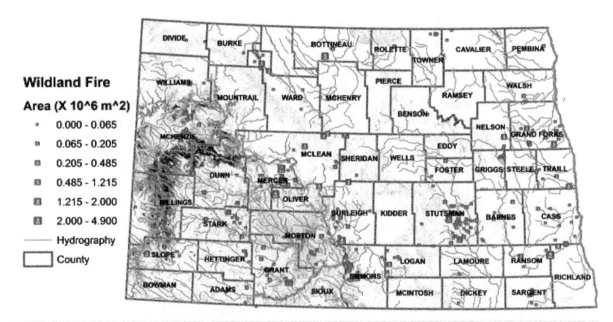

Figure 7. A map depicting 2012 wildland fire locations with graduated symbols depicting burned areas. The hydrography layer shows all perennial, intermittent and ephemeral streams.

2-Grains, hay, seeds, 3-Other crops,4-Idle cropland/fallow/Conservation Reserve Program (CRP), 5-Grass, pasture, non-agriculture, 6-Woodland, 7-Urban/developed, 8-Water. The highest change in area (**Figures 11(a)-(c)**) between 2000 and 2001 is the water category, as shown by the prominent peak. This peak change can be attributed to significant flooding in 2001.Other notable peaks include grains, hay, seeds and idle cropland/ fallow/CRP transiting to grass, pasture, non-agriculture category. This also may be due to more lands converting to CRP lands. The CRP is a land conservation program administered by the Farm Service Agency (FSA) where farmers enrolled in the program agree to remove environmentally sensitive land from agricultural production

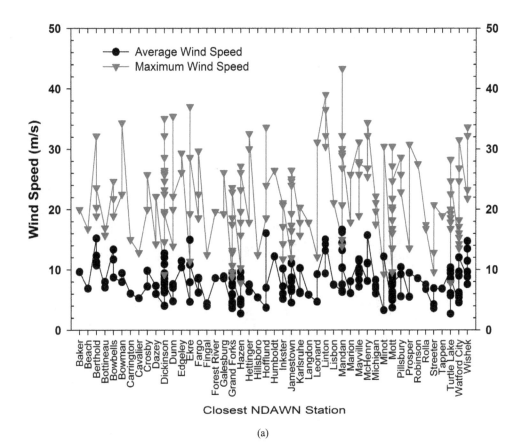

(a)

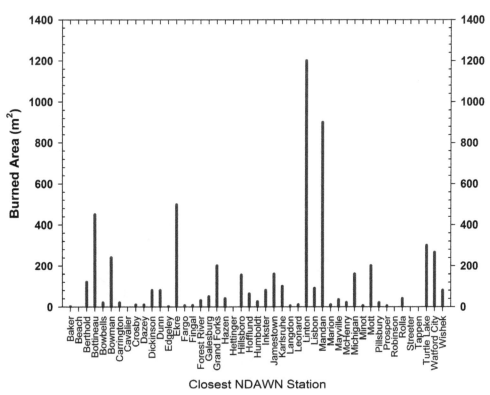

(b)

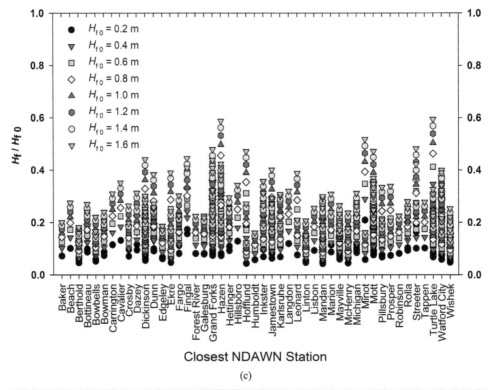

(c)

Figure 8. (a). Average and maximum wind speed data for days when wildland fire was recorded. (b) Burned area within each defined thiessen polygon corresponding to a North Dakota Agricultural Weather Network (NDAWN) station. (c). Ratio of flame height H_f to flame height in the absence of wind H_{f0} for each wildland fire location.

$H_{f0} = 0.20$ m

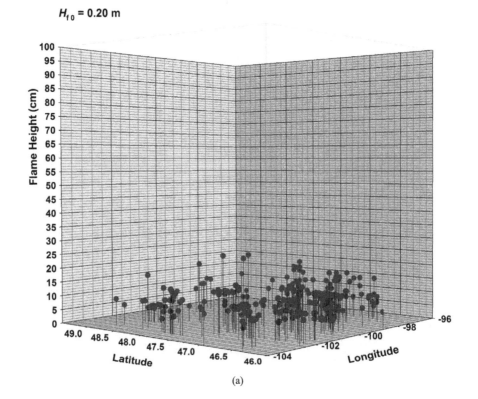

(a)

$H_{f0} = 0.40$ m

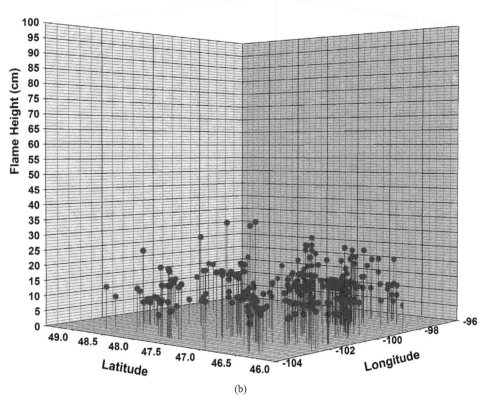

(b)

$H_{f0} = 0.60$ m

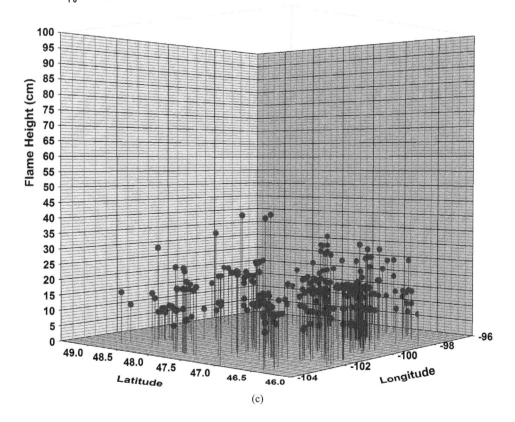

(c)

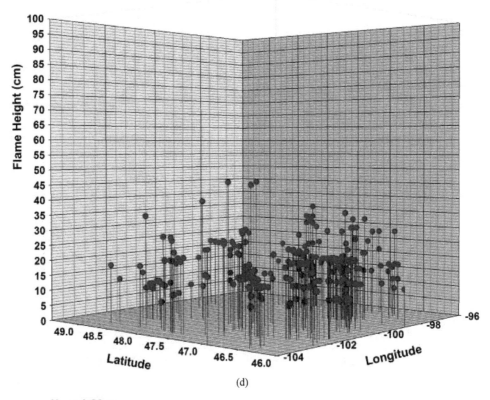

(d)

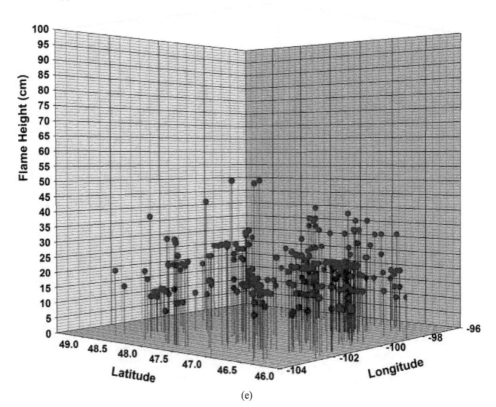

(e)

$H_{f0} = 1.20$ m

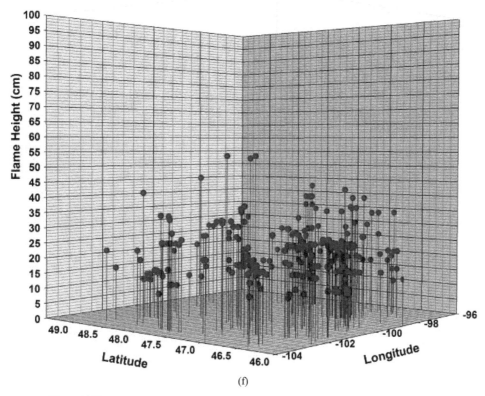

(f)

$H_{f0} = 1.40$ m

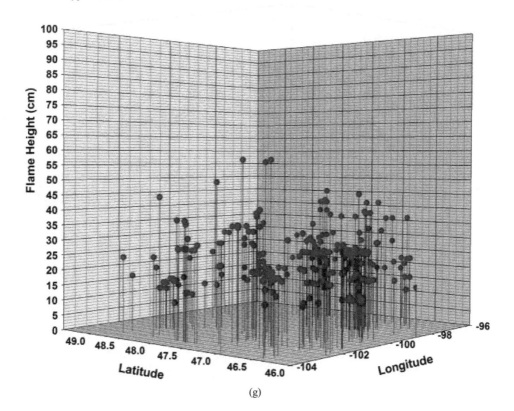

(g)

$H_{f0} = 1.60$ m

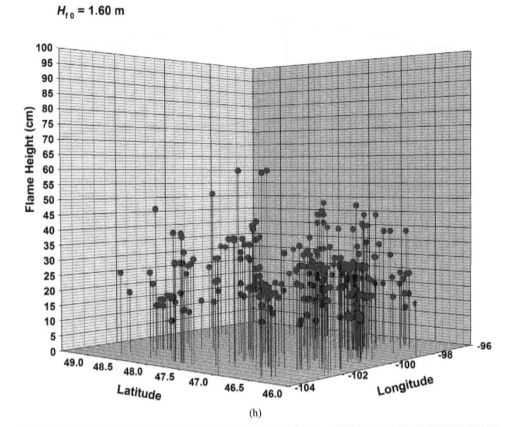

(h)

Figure 9. Graphs of flame heights for various values of H_{f0}.

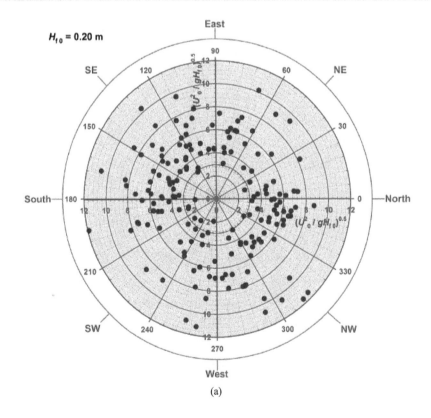

(a)

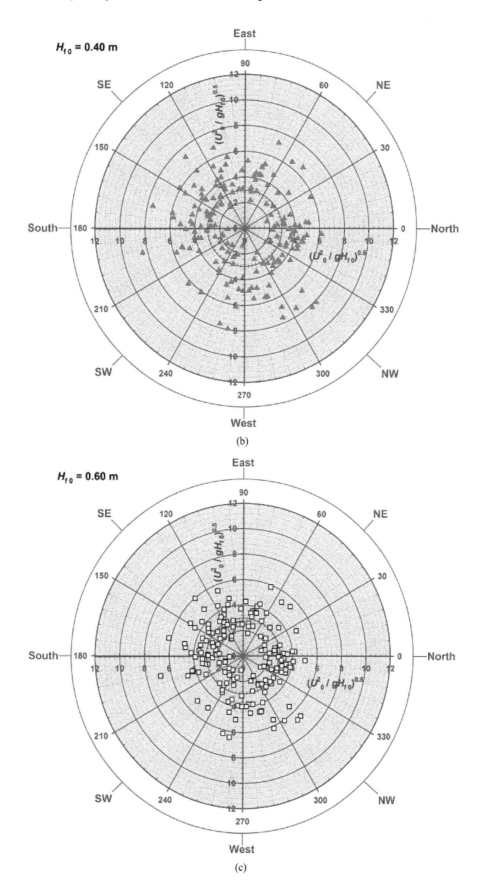

(b)

(c)

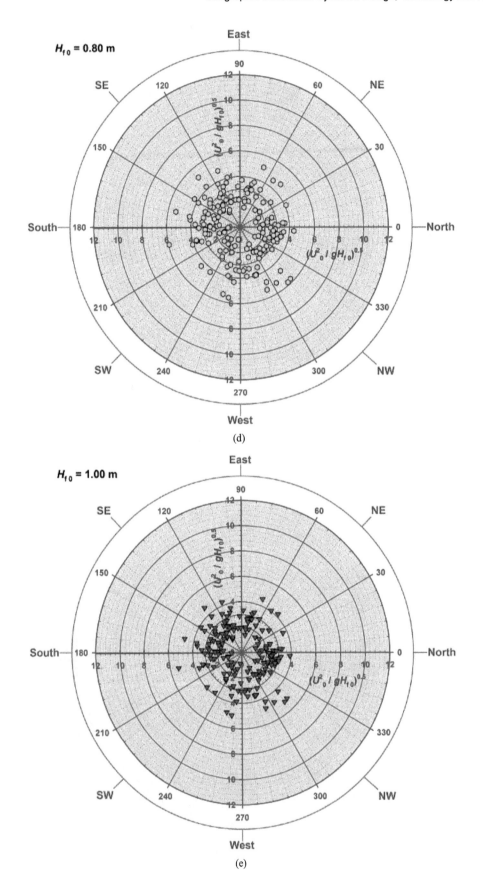

(d)

(e)

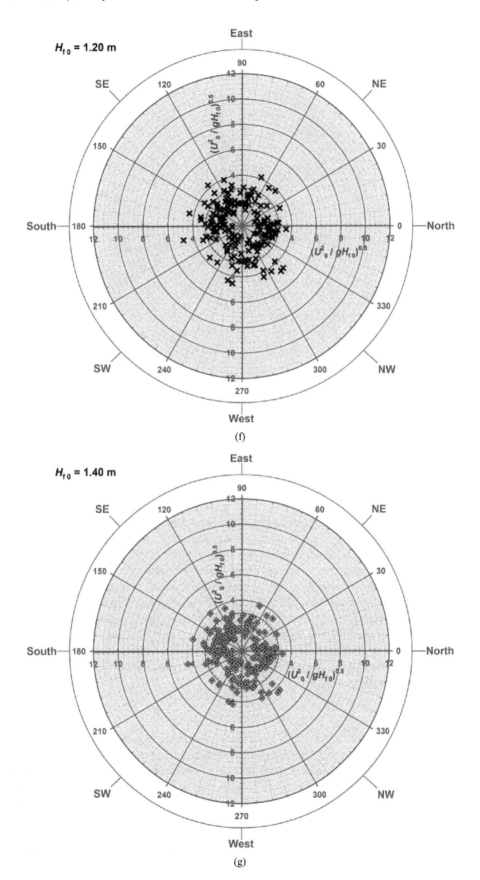

(f)

(g)

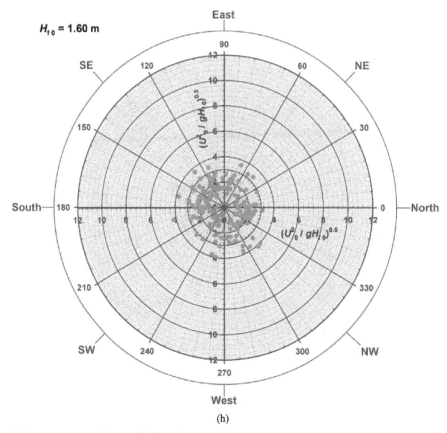

(h)

Figure 10. Polar plots depicting radial variation for $\sqrt{\left(U_0^2/gH_{f0}\right)}$ against prevailing wind direction on the angular axis for various H_{f0} ..

and plant species that will improve environmental health and quality. In the period, 2001-2002 areas that were under floodwaters can be seen transiting to other "absorbing" states, for example categories 2 to 7. Categories that exhibited significant changes between 1999-2001 include other crops, urban/developed areas, and woodland. Considering this subset of the state containing two of the major metropolitan areas, we can assess impacts of forest conversion. For example from the transition probabilities graphs (**Figures 11(d)-(f)**), woodland transition probabilities to urban/developed category are 0.0095 in 1999-2000, 0.0467 in 2000-2001 and 0.0411 for the full period 1999-2001. This trend is low probably due to most development on the eastern side of North Dakota is not significantly affecting predominant riparian forests that exist in this region. But again since this area is greatly affected by periodic flooding, for the time period 2000-2001, there were more instances of water category transiting to other states. Areas with grains, hay, seeds category almost displayed a uniform probability of staying unchanged over the three year period.

5. Conclusions

The goal of our research was to provide a systematic, quantitative and innovative tool that supports decision-makers in forestry management. The designed system can be used to access forest management program initiatives, especially where these programs are lacking. Furthermore, the system is an integral component in spatially displaying areas where the best forestry policies may achieve the best results. The ADSS was designed as an adaptive dynamic framework modularly constructed to optimize system capabilities and provide flexibility for future. The system provides for real-time assessment of information stemming forth from all affiliated entities. The system has a base stage deployed using ArcGIS Web services. Web applications and services utilize-configured authentication methods and a. NET security standard over HTTP with windows security systems recognition.

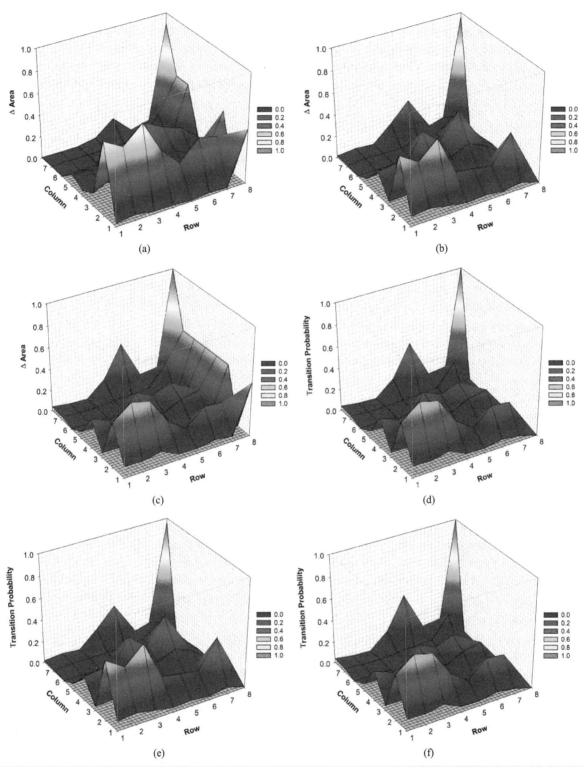

Figure 11. Percent area change for period (a) 1999-2000, (b) 2000-2001, (c) 2001-2002. Transition probabilities for period (d) 1999-2000, (e) 2000-2001, (f) 2001-2002.

From this ADSS a secure gateway for the delivery of North Dakota Spatial Analysis Project (SAP) derived data layers was designed. From this platform, resource information and key management variables can be retrieved, queried or catalogued over non-authenticated web browsers. The ADSS has arisen as a vital link of city, county, state and federal forestry reporting mechanism for NDFS and affiliated partners. For example, in the 2001 year alone, wild land fire reporting peaked at over 930 records marking a significant success. General public can utilize the system by querying multiple layers of records and as such decipher management accomplishments. Since the system also provides baseline data on other resources, for example, watersheds and geo-corrected hydrologic datasets, these can be utilized to secondarily address water quality issues or management opportunities identified through other resources.

Several resource management forms were incorporated into the ADSS, for example, the most intricate form designed was for tracking community accomplishments for urban and community forestry programs. Measurable outcomes include impact of forestry programs on local communities within each fiscal year, impact of professional advice offered to individuals or communities, and federal funding per capita amongst other deliverables. Other forms include wild land fire, innovation in State and Private Forestry, forestry assistance and accountability measures, forestry training programs, forest health, and an online chat. Integral to most of these forms are associated relative databases. The databases store retrievable pertinent information related to forestry resource management. Multi-faceted information can be retrieved from the system from current reports, filed reports or information that can be utilized in a myriad of possibilities.

Acknowledgments

The research reported in this paper was supported in part by US Department of Agriculture Forest Service award # 10-DG-11010000-011 and CFDA Cooperative Forestry Assistance # 10.664, North Dakota Forest Service, and by the North Dakota State University Department of Geosciences. The opinions expressed in this paper are solely those of the writers and are not necessarily consistent with the policies or opinions of the USDA, the USDA-Forest Service, North Dakota Forest Service, North Dakota State University, and/or CFDA.

References

[1] Easterbrook, S. (2003) Model Management and Inconsistency in Software Design. In: Sullivan, K., Ed., *NSF Workshop on the Science of Design*: *Software and Software-Intensive Systems*, NSF, Airlie Center, VA, 2.

[2] Chin, D.N. (1989) KNOME: Modelling What the User Knows in UC. In: Kobsa, A. and Wahlster, W., Eds., *User Models in Dialog Systems*, Berlin Springer-Verlag, 74-107. http://dx.doi.org/10.1007/978-3-642-83230-7_4

[3] Sprague, R.H. and Carlson, E.D. (1982) Building Effective Decision Support Systems. Prentice-Hall International Inc., London, 329pp.

[4] Stock, M.W. and Rauscher, H.M. (1996) Artificial Intelligence and Decision Support in Natural Resource Management. *New Zealand Journal of Forestry Science*, **26**, 145-157.

[5] Chuang, T.T. and Yadav, S.B. (1998) The Development of an Adaptive Decision Support System. *Decision Support Systems*, **24**, 73-87. http://dx.doi.org/10.1016/S0167-9236(98)00065-7

[6] Holsapple, C.W., Pakath, R., Jacob, V.S. and Zaveri, J.S. (1993) Learning by Problem Processors: Adaptive Decision Support Systems. *Decision Support Systems*, **10**, 85-108. http://dx.doi.org/10.1016/0167-9236(93)90032-X

[7] Santos, B.L.D. and Holsapple, C.W. (1989) A Framework for Designing Adaptive DSS Interfaces. *Decision Support Systems*, **5**, 1-11. http://dx.doi.org/10.1016/0167-9236(89)90024-9

[8] Klashner, R. and Sabet, S. (2007) A DSS Design Model for Complex Problems: Lessons from Mission Critical Infrastructure. *Decision Support Systems*, **43**, 990-1013. http://dx.doi.org/10.1016/j.dss.2005.05.027

[9] Krogsaeter, M., Oppermann, R. and Thomas, G.C. (1994) A User Interface Integrating Adaptability and Adaptively, In: Opper-mann R. (Ed.) Adaptive User Support: Ergonomic Design of Manually and Automatically Adaptable Software, Lawrence Erlbaum Associates Publishers, Hillsdale, NJ, 97-125.

[10] Norcio, A.Y. and Staley, J. (1989) Adaptive Human-Computer Interfaces: A Literature Survey and Perspective, IEEE Transactions on Systems. *Man and Cybernetics*, **19**, 399-408. http://dx.doi.org/10.1109/21.31042

[11] Tyler, S.W., Schlossberg, J.L., Gargan, R.A., Cook Jr., L.K. and Sullivan, J.W. (1991) An Intelligent Interface Architecture for Adaptive Interaction, In: Sullivan J.W., Tyler S.W. (Eds.) Intelligent User Interfaces, Addison-Wesley Publishing, Reading, MA, 85-109.

[12] Haugen, D.E., Kangas, M., Crocker, S.J., Perry, C.H., Woodall, C.W., Butler, B.J., Wilson, B.T. and Kaisershot, D.J. (2009) North Dakota's Forests 2005. Resource Bulletin NRS-31. US Department of Agriculture, Forest Service, Northern Research Station, 82pp.

[13] Oduor, P.G., Kotchman, L., Nakamura, A., Jenkins, S. and Ale, G. (2012) Spatially Constrained Forest cover Dynam-

ics Using Markovian Random Processes. *Forest Policy and Economics*, **20**, 36-48.
http://dx.doi.org/10.1016/j.forpol.2012.02.005

[14] Duncan, B.W., Shao, G. and Adrian, F.W. (2009) Delineating a Managed Fire Regime and Exploring Its Relationship to the Natural Fire Regime in East Central Florida, USA: A Remote Sensing and GIS Approach. *Forest Ecology and Management*, **258**, 132-145. http://dx.doi.org/10.1016/j.foreco.2009.03.053

[15] Stambaugh, M.C. and Guyette, R.P. (2008) Predicting Spatio-Temporal Variability in Fire Return Intervals Using a Topographic Roughness Index. *Forest Ecology and Management*, **254**, 463-473.
http://dx.doi.org/10.1016/j.foreco.2007.08.029

[16] Nmira, F., Consalvi, J.L., Boulet, P. and Porterie, B. (2010) Numerical Study of Wind Effects on the Characteristics of Flames from Non-Propagating Vegetation Fires. *Fire Safety Journal*, **45**, 129-141.
http://dx.doi.org/10.1016/j.firesaf.2009.12.004

[17] Kokkinos, E.A. and Maras, A.M. (1997) A First-Order Stationary Markov Class A Transition Density. *Journal of Franklin Institute*, **334**, 525-537. http://dx.doi.org/10.1016/S0016-0032(96)00102-0

[18] Wu, Q., Li, H., Wang, R., Paulussen, J., He, Y., Wang, M., Wang, B. and Wang, Z. (2006) Monitoring and Predicting Land Use Change in Beijing Using Remote Sensing and GIS. *Landscape and Urban Planning*, **78**, 322-333.
http://dx.doi.org/10.1016/j.landurbplan.2005.10.002

[19] Yin, K.K., Yang, H., Daoutidis, P. and Yin, G.G. (2003) Simulation of Population Dynamics Using Continuous-Time Finite-State Markov Chains. *Computers and Chemical Engineering*, **27**, 235-249.
http://dx.doi.org/10.1016/S0098-1354(02)00179-5

Permissions

All chapters in this book were first published by Scientific Research Publishing; hereby published with permission under the Creative Commons Attribution License or equivalent. Every chapter published in this book has been scrutinized by our experts. Their significance has been extensively debated. The topics covered herein carry significant findings which will fuel the growth of the discipline. They may even be implemented as practical applications or may be referred to as a beginning point for another development.

The contributors of this book come from diverse backgrounds, making this book a truly international effort. This book will bring forth new frontiers with its revolutionizing research information and detailed analysis of the nascent developments around the world.

We would like to thank all the contributing authors for lending their expertise to make the book truly unique. They have played a crucial role in the development of this book. Without their invaluable contributions this book wouldn't have been possible. They have made vital efforts to compile up to date information on the varied aspects of this subject to make this book a valuable addition to the collection of many professionals and students.

This book was conceptualized with the vision of imparting up-to-date information and advanced data in this field. To ensure the same, a matchless editorial board was set up. Every individual on the board went through rigorous rounds of assessment to prove their worth. After which they invested a large part of their time researching and compiling the most relevant data for our readers.

The editorial board has been involved in producing this book since its inception. They have spent rigorous hours researching and exploring the diverse topics which have resulted in the successful publishing of this book. They have passed on their knowledge of decades through this book. To expedite this challenging task, the publisher supported the team at every step. A small team of assistant editors was also appointed to further simplify the editing procedure and attain best results for the readers.

Apart from the editorial board, the designing team has also invested a significant amount of their time in understanding the subject and creating the most relevant covers. They scrutinized every image to scout for the most suitable representation of the subject and create an appropriate cover for the book.

The publishing team has been an ardent support to the editorial, designing and production team. Their endless efforts to recruit the best for this project, has resulted in the accomplishment of this book. They are a veteran in the field of academics and their pool of knowledge is as vast as their experience in printing. Their expertise and guidance has proved useful at every step. Their uncompromising quality standards have made this book an exceptional effort. Their encouragement from time to time has been an inspiration for everyone.

The publisher and the editorial board hope that this book will prove to be a valuable piece of knowledge for researchers, students, practitioners and scholars across the globe.

List of Contributors

Omar Bachir Alami and Fatima Bardellile
Ecole Hassania des Travaux Publics, Casablanca, Maroc

Hatim Lechgar and Mohamed El Imame Malaainine
Faculté des Sciences Ain Chock, Université Hassan II, Casablanca, Maroc

Hala A. Effat
Department of Environmental Studies and Land Use, National Authority for Remote Sensing and Space Sciences (NARSS), Cairo, Egypt

Arjun Raj Pandey
Department of Civil, Architectural, and Environmental, Illinois Institute of Technology, Chicago, USA

Farzad Shahbodaghlou
Departmant of Engineering, California State University, East Bay, Hayward, USA

Chidinma Blessing Okoye
Department of Geography, University of Lagos, Akoka, Nigeria

Vincent Nduka Ojeh
Wascal, Department of Meteorology, Federal University of Technology, Akure, Nigeria

Kadeghe G. Fue and Camilius Sanga
Computer Centre, Sokoine University of Agriculture, Morogoro, Tanzania

Asma Belasri and Abdellah Lakhouili
Faculty of Science and Technology, Hassan 1 University, Settat, Morocco

Ioannis Pispidikis and Efi Dimopoulou
School of Rural and Surveying Engineer, National Technical University of Athens, Athens, Greece

Silvia Maria Santana Mapa
Federal Office for Education, Science and Technology of Minas Gerais, Congonhas, Brazil

Renato da Silva Lima
Federal University of Itajuba, Industrial Engineering and Management Institute, Itajubá, Brazil

Khalid A. Eldrandaly, Soaad M. Naguib and Mohammed M. Hassan
Information Systems Department, Faculty of Computers and Informatics, Zagazig University, Al-Sharqiyah, Egypt

Hicham Lahlaoi, Hassan Rhinane, Atika Hilali and Loubna Khalile
Geosciences Laboratory, Faculty of Sciences Ain Chock, Hassan II University, Casablanca, Morocco

Said Lahssini
National School of Forestry Engineering, Salé, Morocco

Shawn Dias, Chaowei Yang, Anthony Stefanidis and Mathew Rice
Department of Geography and GeoInformation Science, George Mason University, Fairfax, VA, USA

Peter G. Oduor
Department of Geosciences, North Dakota State University, Fargo, USA

Michael Armstrong
IT Systems Vulnerability, Sanford Health Fargo, Fargo, USA

Larry Kotchman
North Dakota Forest Service, Bottineau, USA

Michael Kangas
North Dakota Forest Service, North Dakota State University, Fargo, USA

Buddhika Maddurapperuma
Environmental & Conservation Science Program, North Dakota State University, Fargo, USA

Kelsey Forward
Wenck Associates North Dakota, Fargo, USA

Pubudu Wijeyaratne
Department of Computer Science, North Dakota State University, Fargo, USA

Xiana Santos
Department of Forestry, Mississippi State University, Starkville, USA

Akiko Nakamura
East View Geospatial, Minneapolis, USA

Krystal Leidholm
California Department of Fish & Game, Washburn, USA

Printed in the USA
CPSIA information can be obtained
at www.ICGtesting.com
JSHW051445221024
72173JS00006B/1589